TEN PLANTS THAT CHANGED MINNESOTA

TEN PLANTS *that changed* MINNESOTA

Mary Hockenberry Meyer and Susan Davis Price

For information, write to the Minnesota Historical Society Press,
345 Kellogg Blvd. W., St. Paul, MN 55102-1906.

Watercolor images at the opening of each chapter by Cathy Gilchrist Kaudy.

www.mnhspress.org

The Minnesota Historical Society Press is a member
of the Association of American University Presses.

Manufactured in the United States of America

10 9 8 7 6 5 4 3 2 1

∞ The paper used in this publication meets the minimum requirements
of the American National Standard for Information Sciences—
Permanence for Printed Library Materials, ANSI Z39.48-1984.

International Standard Book Number
ISBN: 978-1-68134-034-0 (paper)

Library of Congress Cataloging-in-Publication Data
Names: Meyer, Mary Hockenberry, author. | Price, Susan Davis, author.
Title: Ten plants that changed Minnesota /
Mary Hockenberry Meyer and Susan Davis Price.
Description: St. Paul, MN : Minnesota Historical Society Press, [2017] |
Includes bibliographical references and index.
Identifiers: LCCN 2016054089 | ISBN 9781681340340 (pbk. : alk. paper)
Subjects: LCSH: Plants—Minnesota.
Classification: LCC QK168 .M49 2017 | DDC 581.9776—dc23
LC record available at https://lccn.loc.gov/2016054089

CONTENTS

We dedicate this book to our grandchildren—
Michael Julio, Rebeca, Merav, Charlotte, Ben, Henry,
Roland, Madeline, Harrison, and Baby Maloney—
with the wish that Minnesota's future and their own
be filled with healthy plants and a healthy planet.
MARY *and* SUSAN

FOREWORD

ARNE CARLSON

It is not necessary to be a botanist to realize how much plants have contributed to our unique quality of life in Minnesota. White pine adds to the beauty of the north woods, American elms stand along Victory Memorial Parkway in Minneapolis, wild rice grows in hundreds of our northern lakes, and corn and soybeans support the northwestern, central, and southern Minnesota economy.

These and the other five plants highlighted by the 10 Plants project help to define our state's splendid landscape.

Indeed, that landscape nurtured Dakota people for millennia and has drawn others here in the last few centuries—from the Ojibwe who sought the food that grows on water (wild rice), to the loggers who harvested white pine to build the West, to the millers who ground flour from the finest wheat grown anywhere in the world.

Today, Minnesota farmers harvest hardy northern strains of soil-building alfalfa that feed extensive dairy herds, and they are near the top in global production of soybeans that make their way to China, Mexico, and Indonesia. Can we imagine what our state would be like without our world-famous Honeycrisp apples, the abundant fields of corn, and the vibrant green turf that covers our golf courses and college campuses?

Perhaps the finest gift we can give to honor those people, past and present, who have embraced the bounty of our native landscape is to have their legacy and their wisdom passed on as an integral part of the future. *Ten Plants That Changed Minnesota* does just that by telling the story of these plants while helping us to see their importance in our world today and in the days to come.

Minnesotans have worked collaboratively to manage invasive purple loosestrife by using biological controls that have restored our wetlands with minimal pesticides. As we learn about the environmental challenges of dams and sulfites encroaching on wild rice, of nitrogen fertilizer used to grow corn, of automated irrigation systems for lawns, we realize the importance of investing in multiple tactics to balance economic growth with environmental preservation, in order that future generations may enjoy these landscapes as well.

Having grown up in a major city, I certainly appreciate the abundance of natural spaces in our state. We have worked to protect Minnesota's wetlands, its prairies, and its "up north" woods, and through books such as this one we can continue to spread the word about the vital importance of our natural areas. What better way to celebrate Minnesota's unique quality of life than by knowing and nurturing the landscapes that define and support us?

INTRODUCTION

MARY HOCKENBERRY MEYER

Plants have always fascinated me. When I learned about plant blindness, the inability to see or notice the plants in one's own environment, I realized why other people were not as excited about plants: they simply did not see them!

Kids today spend an average of fifty-three hours a week in front of a screen rather than outside wandering on a farm or in a garden, the woods, or a natural area, as I did growing up. Are kids today spending even one hour a week looking at plants? More than one generation removed from growing up on a farm, young people may look at a beet in the grocery store and ask, "What *is* that?" If we cannot name and recognize plants, how can we value them and realize how essential they are to our environment and our well-being as humans?

The idea that plants, as few as ten, could shape a state and how it developed economically, culturally, and historically enticed me to develop the 10 Plants That Changed Minnesota project.

What are the 10 Plants? How could they be selected?

A public nomination process launched and publicized by the Minnesota Landscape Arboretum in February 2012 started the 10 Plants project. That spring, more than 500 Minnesotans nominated plants they thought should be on the top ten list. You can read their comments in the sidebars throughout this book. Plants you would expect, as well as dandelions, Cuties mandarin oranges (!), buckthorn, and creeping Charlie, all made the list. The merits of these plants were evaluated based on their impact—both positive and negative—in six areas: environmental; economic or industrial; cultural/spiritual; historical; sustenance; and landscape. The final 10 Plants were judged and selected by Alan Ek, professor, University of Minnesota Department of Forest Resources; Al Withers, director,

Minnesota Agriculture in the Classroom; Beverly Durgan, dean, University of Minnesota Extension; Bob Quist, director, Oliver Kelley Farm/Minnesota Historical Society; Brian Buhr, dean, University of Minnesota College of Food, Agricultural and Natural Resource Sciences; Karen Kaler, University of Minnesota President's Office; Mary Maguire Lerman, president, Minnesota State Horticultural Society; Nancy Jo Ehlke, professor, University of Minnesota Department of Agronomy and Plant Genetics; Susan Bachman West, owner, Bachman's Inc.; Gary Gardner, professor, University of Minnesota Department of Horticultural Science; Neil Anderson, professor, University of Minnesota Department of Horticultural Science; Karl Foord, University of Minnesota Extension educator; and Mary Meyer, professor, University of Minnesota Department of Horticultural Science.

Since 2012, a website, a University of Minnesota freshman seminar, a kids contest for games and activities associated with the 10 Plants, and now this book and the separate teachers activities handbook for use with middle and high school students have been developed to inspire and educate Minnesotans, especially young people. Although the 10 Plants That Changed Minnesota project was my idea, I was lucky to know Susan Davis Price, an accomplished author, who graciously agreed to write the vast majority of the book and the text for the original website. Susan has written each chapter, compiled the historical information, and selected many of the images. Susan wrote the book; I had the idea and helped with editing and arranging for academic experts to review the chapters.

Our objective for the book is to inform Minnesotans about the importance of plants in our lives and how these specific plants helped our state become what it is today. We hope to inspire young people to reflect on the past and think about what plants will impact the future. Due to the geographic and climatic diversity of the state, almost every county in Minnesota has been influenced differently by the 10 Plants: from the wild rice lakes and oldest white pine stands of the north, to the metropolitan parks and golf courses, to the endless fields of corn and soybeans in the south. The huge acreage that many of these plants cover has significantly affected the environment. Some, such as corn, turfgrass, and soybeans, have greatly increased in acreage; others, such as white pine and wild rice, have been reduced. The American elm once provided a cathedral of branches for urban streets, and purple loosestrife invaded wetlands.

Minnesota's geographic location near the center of a large land mass, with no major water body to buffer temperatures and precipitation reduction from the forests of the east to the dryer grasslands in the west, resulted in diverse vegetative regions that not only dictated what plants grew here originally but also determined what plants we grow today. Corn, humans' domesticated grass, has replaced the tallgrass prairie, or sea of grass, the pioneers saw atop some of the richest, most productive soil in the world. Simplistically, Minnesota's plant communities are the prairie in the south, west, and northwest; the coniferous forest in the northeast; and the deciduous forest in the southeast. Within these three major biomes are distinct regions of specific plant communities. Working from original surveyors' written notes, Francis J. Marschner created a map in less than three years, having never set foot in Minnesota. Called the Marschner Map of Original Vegetation, it is an amazing picture of vegetation nearly 200 years ago (see pages 4–5).

Underlying vegetative communities is an equally diverse and critically important feature of soil, which plays a major role in what plants grow throughout the state. It is easy to see the similarities between plant communities and soil types when comparing two maps, the Marschner map and the University of Minnesota soils map (see page 6). The deep organic soils of the prairie in the west supported grasslands while the thin mineral soil of the northeast favored forests. Throughout the book, readers will see the importance of soil and vegetative community in what plants are grown in these regions today.

Ten Plants That Changed Minnesota highlights the importance of balancing the choices we have made in growing and managing these plants. Harvesting centuries-old white pine to build our cities and converting thousands of acres of prairie to grow tilled monocultures

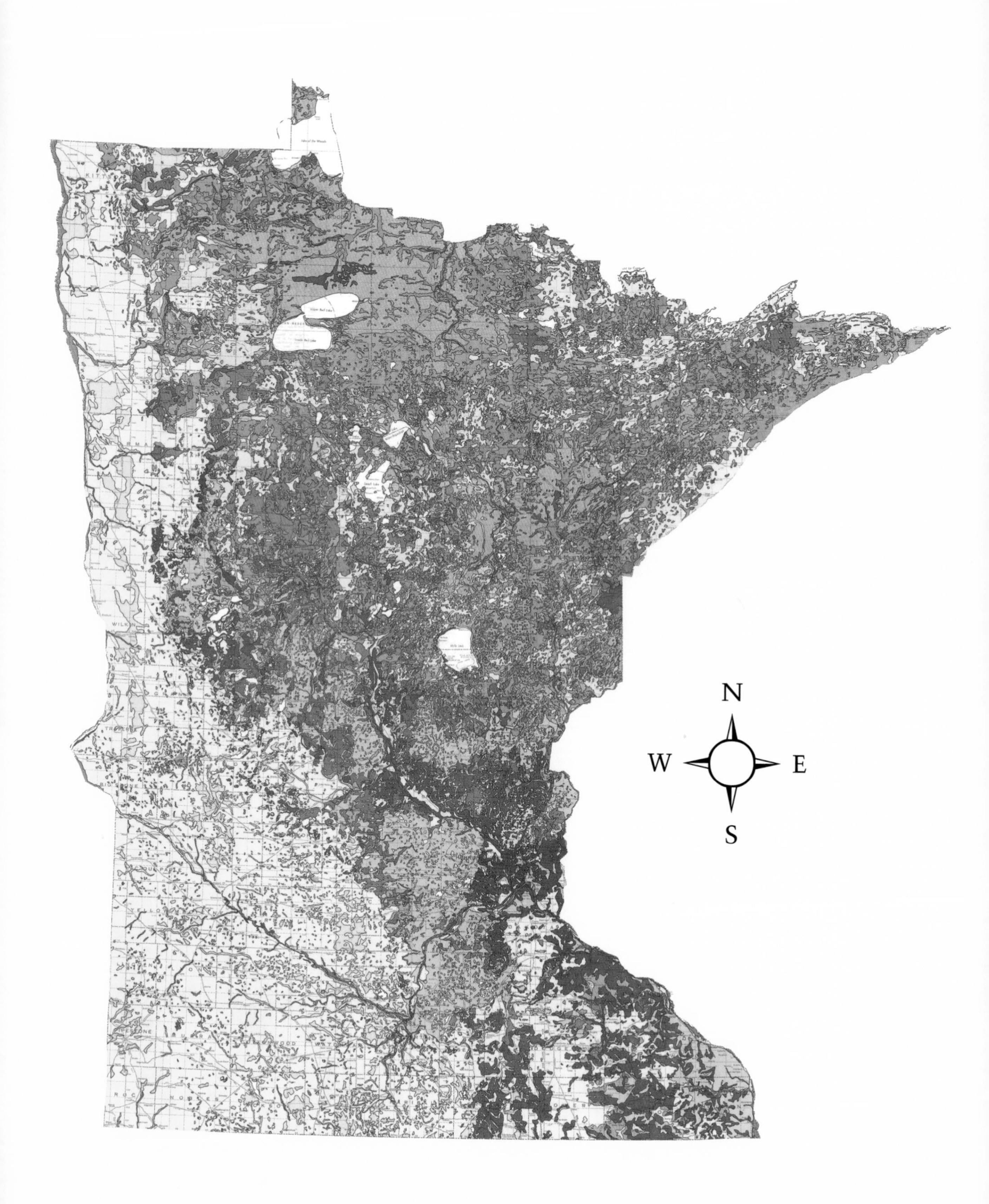
N
W
E
S

THE ORIGINAL VEGETATION OF MINNESOTA

Scale: 1:500,000

Miles

0 10 20 30 40

LEGEND

GRASSLAND

Prairie

WET PRAIRIES, MARSHES, AND SLOUGHS Marsh-grasses, flags, reeds, rushes, wild rice, with willow and alder-brush in places.

BRUSHLAND

BRUSH PRAIRIE Grass and brush of aspen, balm of Gilead, and little oak and hazel in the north but mainly oak and hazel in the south.

ASPEN-OAK LAND Aspen, generally dense, but small in most places with scattered oaks and few elms, ash, and basswood.

OAK OPENINGS AND BARRENS Scattered trees and groves of oak (mostly bur oak) of scrubby form with some brush and thickets and occasionally with pines.

HARDWOOD FOREST

BIG WOODS Oaks (bur, white, red, and black), elm, basswood, ash, maple, hornbeam, aspen, birch, wild cherry, hickory, butternut, black walnut, etc., with some white pine.

RIVER-BOTTOM FOREST Elm, ash, cottonwood, boxelder, oaks, basswood, soft maple, willow, aspen, hackberry, etc., with occasional pines and arborvitae in the pine region.

ASPEN-BIRCH (HARDWOODS) Will eventually become hardwoods. Includes ash, elm, maple, basswood, oaks, etc., as associated species.

PINERIES

MIXED HARDWOOD AND PINE Maple, white pine, basswood, oaks, hornbeam, ash, elm, aspen, birch, and balsam fir in the north.

PINE GROVES Nearly pure stands of pine.

White pine.

White and Norway pine.

JACK PINE BARRENS AND OPENINGS Jack pine with oak, aspen, hazel-brush, and occasionally Norway pine.

PINE FLATS Hemlock, spruce, fir, cedar, and white pine.

ASPEN-BIRCH (CONIFER) Will eventually become conifer. Includes white and Norway pines, balsam, fir, birch, spruce, and arborvitae as associated species.

BOGS AND SWAMPS

CONIFER BOGS AND SWAMPS Spruce, tamarack, cedar, and balsam.

OPEN MUSKEG (FLOATING BOGS) Mosses, rushes, marsh-grasses, alder-brush, scattered small tamarack.

Francis Marschner's map of the original vegetation of Minnesota was created from the 200-volume Public Land Survey, 1847–1907. MNHS collections

SOIL SUBORDERS OF MINNESOTA

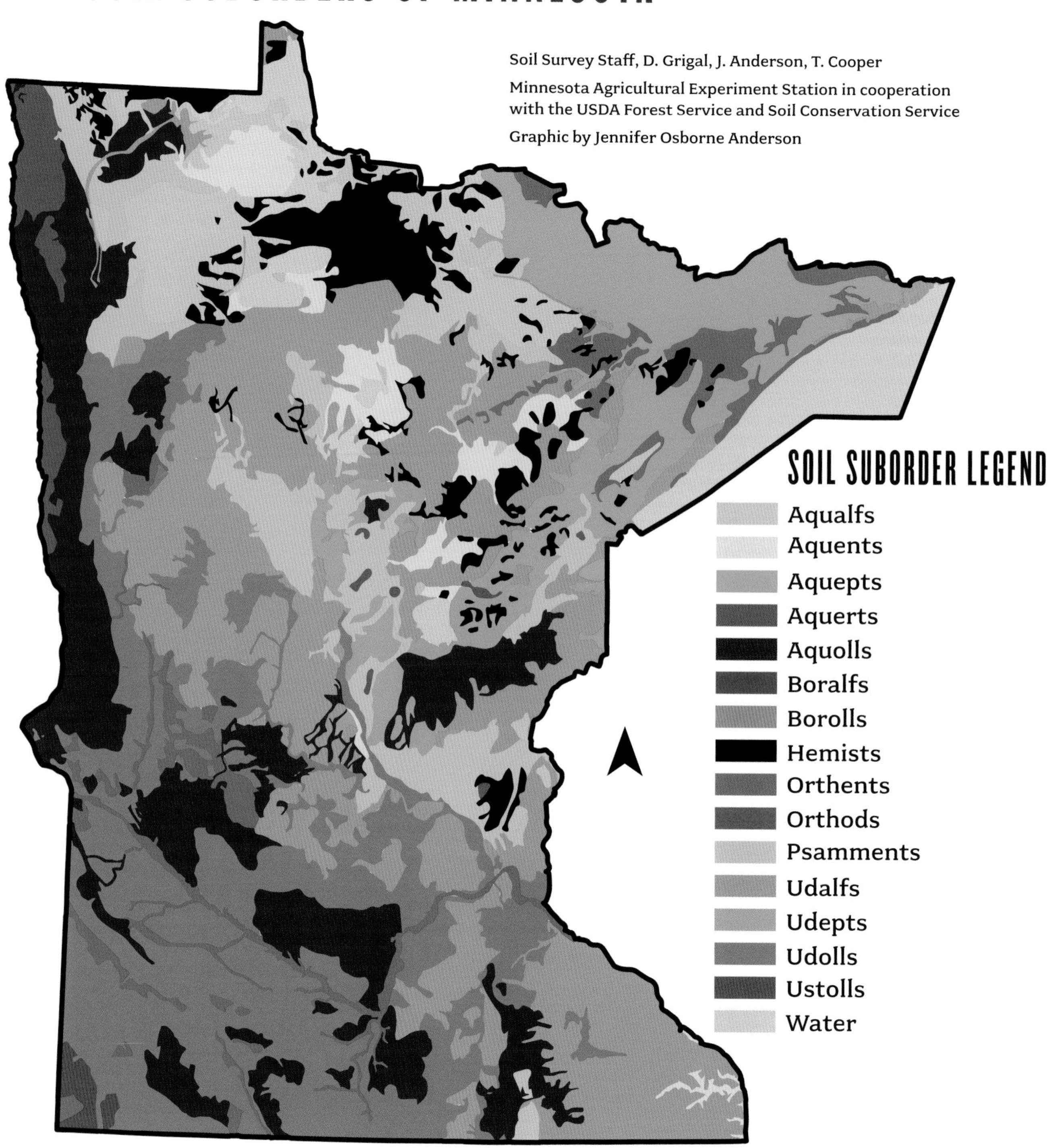

have negatively impacted the environment while boosting the economy for the state. Should economics always be the winner? In the teachers activities handbook, we offer several opportunities for students to debate these choices. The decisions are not easy, but we all share the responsibility of being informed, especially when the results affect all of us.

This book highlights only ten plants, but we hope that greater awareness of these plants will help reduce plant blindness and increase our ability to *see* plants more generally. We know plants provide our food, directly and indirectly, but they also provide the grass carpet for recreational games, activities, rest, and relaxation. Plants provide natural restorative settings for relief from technology overload and peace and quiet for reflection and renewal. Beautiful plants surrounding us can positively affect our minds, our emotions, and our physical health. A walk in the woods or in a beautiful garden restores our sense of calm and reduces stress. Nurturing plants in a garden or even caring for lawns, apple trees, or white pines has proved to have mental and physical benefits, relieving symptoms of anxiety and lowering blood pressure.

We hope this book and the accompanying teachers activities handbook will inspire young people and adults to look at plants, to grow plants, and to think about how we manipulate plants to benefit us and to feed the world. While plant-related careers are not a specific topic of the book, students will easily see the many rewarding opportunities available in horticulture, agriculture, conservation, forestry, ecology, and plant and food science that are vital for us to feed the world's population and maintain the earth for the next generation. Our goal is for all Minnesotans, no matter what their age or the county in which they live, to see these plants with new eyes and appreciate how they have shaped our state and impacted each of us.

Soil suborders of Minnesota shows the wide variation of soil types across the state.

INTRODUCTION

SUSAN DAVIS PRICE

My initial knowledge of the 10 Plants was a superficial one, and thus I assumed the task of writing this book would be a simple one. Fortunately, I was soon proved wrong and in the process uncovered the rich history of these plants. Just as first impressions of friends often are replaced in time by more nuanced and layered understandings of their character, so my knowledge of these plants became more complex with deeper research. Corn is not only that ear we happily munch in summer; purple loosestrife is not simply that tall, pretty plant we see along our waterways; wheat is not just an ingredient in our bread. All these plants have entangled histories, uses, benefits, and associated hazards.

In conducting the research for this book, I came across surprising facts, such as the antiquity of some of the 10 Plants. I was fascinated to learn that white pine's appearance in Minnesota preceded human habitation by thousands of years, with the trees moving into the area just after the last Ice Age, about 7,000 years ago. Corn, indigenous to the Western Hemisphere, is ancient as well. Corn pollen grain, thought to be 10,000 years old, has been obtained from drill cores 200 feet below Mexico City. Wheat, too, has a long history. Historians report that about 17,000 years ago people chewed on the wild kernels and so discovered their tastiness and nutritive qualities.

Many of the plants have been used as medicine. White pine resin contains a number of antimicrobials that the Ojibwe used to successfully treat infections and even gangrene. Purple loosestrife has astringent and diuretic properties and can be applied to the skin for problems such as eczema and dermatitis. A homeopathic tincture has been made of the inner bark of elm and used as an astringent. Currently, Americans may say, "An apple a day keeps the doctor away," but herbalists in earlier times listed more than fifteen ways that apples could be used medicinally. The least likely current application: rotten apples were used as a poultice for sore eyes in long-ago America.

The plants themselves are fascinating, but the people involved are even more so. How could we not be amazed at the talent among us—that the University of Minnesota scientist David Bedford could taste hundreds of apples in a summer and still discover a Honeycrisp, a Zestar! We must admire the tenacity of the German immigrant Wendelin Grimm, who brought alfalfa seeds with him in 1857, planted them out, and saved the hardiest seeds every year. After many years of this practice, he had created a winter-hardy stock, one that became the basis of all the alfalfa grown in the United States for more than a hundred years.

In 1981 Dr. James Orf, after a stint in the Peace Corps, came to the university and was instrumental in developing fifty general-purpose varieties as well as sixty special-purpose varieties of soybeans. His work has produced soybeans that can resist disease, produce greater yields, and thrive in northerly climates.

Who would have suspected the rich intrigue behind the development of wheat-processing plants? Millers kept their coveted processes secret, to maintain a corner on the market. To counter the secrecy, some adopted desperate measures. Even the prominent Charles A. Pillsbury was known to have "pocketed" wheat seeds at his competitor's mill, in hopes of discovering that miller's reasons for success.

In another instance, flour buyer George H. Christian, in visit after visit, quietly observed the Dundas mill, which was known for its superior product. Over time, Christian ferreted out its processes and took the information to Cadwallader Washburn, owner of the Washburn B Mill in Minneapolis. Washburn hired Christian as the new manager, and under his direction, the Washburn B became one of Minnesota's preeminent mills, with production increasing from 600 barrels of flour a day to 50,000.

In the broader sense, this history tells us how much these specific plants have impacted Minnesota, providing building materials for generations, contributing livelihoods and nutrition to the indigenous people and immigrants, adding beauty to our landscapes, helping make great cities on the plains. In addition, Minnesotans have changed the plants—introduced them, improved them, reduced their numbers. Utilizing these plants is not a neutral activity but one that has consequences. We hope with this book to create a renewed appreciation of the integral role plants have in our politics, our economies, our environments, and our quality of life.

COOL FACTS *about the* 10 PLANTS

WHITE PINE was the Tree of Peace to the Iroquois Nation; its cluster of five needles signified the unification of the Five Nations: Mohawk, Oneida, Onondaga, Cayuga, and Seneca.

WILD RICE is a grain that expands when cooked; it can also be popped, similar to popcorn.

One acre of **SOYBEANS** can be made into more than 82,000 crayons.

ALFALFA can grow roots forty-nine feet long.

Approximately fifty **APPLE** leaves are needed to produce enough energy for one apple.

Seventy-five percent of all grocery items contain some form of **CORN**.

A fifty-by-fifty-foot **LAWN** releases enough oxygen throughout the day to support the needs of one person.

WHEAT contains more protein than corn.

Hockey sticks made from **AMERICAN ELM** bend without splitting.

Popsicle sticks made from **WHITE PINE** are soft and nontoxic and do not splinter.

Honeybees love **PURPLE LOOSESTRIFE** and make delicious honey from this plant.

1
ALFALFA

We have seen the green alfalfa
As it lay beneath the sun,
We have seen it dried and gathered
When the harvest time is done

SUNG TO THE "THE BATTLE HYMN OF THE REPUBLIC"
AT ALFALFA DAY, THIEF RIVER FALLS, 1925

Alfalfa, often called "Queen of Forages," is one of the most important crops in the United States, due to its high yield, wide adaptation, disease resistance, excellent feeding quality, and soil benefits. It makes a tremendous contribution to Minnesota's agricultural economy, to the US economy, and to food production, but it often goes unrecognized. It's like the restrained, subdued classmate beside the flashy, noticeable kids (corn and soybeans).

Indeed, alfalfa made Minnesota's dairy industry possible. Because of its superior protein content and digestible fiber, it has often been used as feed for high-producing dairy cows. Almost 150 years ago, Minnesota dairying led to the establishment of the first farm cooperative in the country. Now Minnesota is home to two of the largest cooperatives in America, Cenex and Land O'Lakes.

In addition, alfalfa is the preferred food for horses, sheep, and beef cattle because it is a good source of vitamins and minerals as well as protein. Because of its high biomass production, perennial nature, and ability to provide its own nitrogen fertilizer, alfalfa has considerable potential in the production of ethanol and other industrial materials.

CULTIVATION HISTORY

Alfalfa (*Medicago sativa*) has an ancient history. Charred remains of the plant, more than 6,000 years old, have been found in southwestern Iran. It seems to have been domesticated near present-day Turkmenistan, Iran, Turkey, the Caucasus region, and other countries in

Alfalfa field. David L. Hansen, University of Minnesota.

Asia Minor and was used to feed the grazing animals that humans had domesticated.

Alfalfa first appeared in civilization's written record in the cuneiform tablets of the Hittites, a people in Asia Minor, around 1300 BCE. A thriving, powerful people, the Hittites wrote that alfalfa was a nutritious plant, one used to feed their horses. The Hittites had invented the war chariot, using it to conquer Egypt and Mesopotamia around 1800 BCE. And alfalfa, with its high yield and high feeding value, gave them strong horses. Some authorities postulate that alfalfa was the critical factor that enabled the horse-driven powers of the region to expand at that time and conquer neighboring rival principalities.

Later, about 700 BCE, alfalfa was included in the list of garden plants of the Babylonian king Merodoch-Baladan. Like other crops, alfalfa journeyed with invading armies and with trade. For example, when the Persians under Xerxes invaded Greece in 480 BCE, they planted fields of alfalfa as fodder for their chariot horses, camels, and other livestock. As the armies retreated, the Greeks saw alfalfa for the first time and quickly adopted it.

The Roman naturalist and philosopher Pliny the Elder (23–79 CE) in his *Natural History* recorded the discovery as follows: "Lucerne [alfalfa] is foreign even to Greece, having been imported from Media during the Persian invasions under Darius; but so great a bounty deserves mention even among the first of the grains, since from a single sowing it will last more than thirty years."

In the second century BCE, the Romans adopted alfalfa from the Greeks. The new forage crop thrived and quickly spread throughout Italy. As the Roman Empire advanced, alfalfa traveled with it. Farmers in the newly acquired regions adopted the practice of growing alfalfa, planting fields of the legume, and so it moved into North Africa, Spain, France, and the Lake Lucerne area of Switzerland. The alternate name, lucerne, seems to have come from the Latin *lucerna*, meaning "lamp," alluding to the plant's bright seeds.

WHAT MINNESOTANS HAD TO SAY ABOUT ALFALFA

Alfalfa is the high-nitrogen crop that is the backbone of Minnesota's dairy and meat industries. A legume, it enriches the soil and thrives without irrigation. The value of the crop can be seen in that while Minnesota is fifth in the United States for production, two-thirds of the growers use all the alfalfa hay they produce. It's too valuable to sell. Its high nitrogen content is the reason our dairy cows can produce quality milk and cream all winter. Alfalfa hay, silage, and chop are leading components in sound farm management.

EMILY Q.

Julianna and Wendelin Grimm, about 1870.
MNHS collections

HOW ALFALFA CAME TO MINNESOTA

In 1857 Wendelin Grimm and his wife, Julianna, and children emigrated from a small farming village in southern Germany and made their home in Carver County. A farmer, Grimm brought with him a twenty-pound bag of *ewiger Klee* (German for "everlasting clover") seeds. *Ewiger Klee*, known to us as alfalfa, had done well in his home country and had been the forage choice for European dairy herds. During the family's first winter in Minnesota, most of the alfalfa was killed by the harsh conditions. Grimm collected seed from the surviving plants and planted them again. A determined man, he continued this practice for a number of years, eventually creating a strain that could reliably survive the frigid winter.

Grimm made no effort to promote his seeds, but his neighbors began to notice his success. One oft-told story recalls a day in 1863 when Grimm herded several of his fat cattle past farmer Henry Gerdsen's place. Observing the cows' healthy girth in contrast to his own lean, corn-fed animals, Gerdsen asked where Grimm had obtained his corn. Grimm replied proudly, "*Kein Kornchen. Nur ewiger Klee*"; "Not one little kernel of corn. Only everlasting clover."

As local farmers observed his cattle and crop success, Grimm shared his seed with them. They began to depend on "Grimm" alfalfa for its reliability and nutritional value. During Grimm's lifetime, his seeds had spread only within a ten-mile radius. At his death in 1890, his obituary made no mention of his significance. However, his sons continued farming with their father's seeds.

Arthur B. Lyman of Excelsior was the first person to take an interest in introducing Grimm's alfalfa to a larger public. Seeing its superiority in Minnesota weather, he brought the story to Professor Willet Hays, head of the University of Minnesota Agricultural Experiment Station. In 1900, with horse and buggy, Hays drove the twenty-five miles with Andrew Boss, an agriculture professor at the university, to the farm, made a detailed inspection of the crops, and returned enthusiastic about the possibilities of promoting hardy alfalfa as a forage crop for the north. After a series of trials at the experiment station, Professor Hays expressed publicly his excitement about the strain, giving it the first official recognition as "Grimm" alfalfa.

In 1904 the US Department of Agriculture (USDA) became actively interested in testing Grimm's alfalfa. In one interesting experiment, the USDA planted seed from Grimm's farm next to fields of alfalfa seed collected from the region of Grimm's old home in Germany. The USDA report, "Wendelin Grimm and Alfalfa," concluded as follows:

> In comparative experiments with Grimm alfalfa and the old German Franconian alfalfa the latter has proved to be much less hardy under our northern conditions than the Grimm, which points to the probability that the German lucern [*sic*] that Mr Grimm brought with him has been greatly modified during its fifty years' sojourn in Minnesota.

> ... Under identical conditions and with identical treatment, adjoining rows of these strains killed out differently. In the old German strain 84 out of 85 plants winterkilled, while in a sample of Grimm grown in North Dakota only 2 out of 70 plants were killed.

Several other northern experiment stations—in North Dakota, Montana, and Kansas—planted Grimm alfalfa seed alongside other northern strains. Grimm's seed consistently produced higher yields and had better winter survival. Grimm grew his alfalfa on upland sites with Lester soil, Minnesota's state soil, known for being well drained and found in sixteen south-central counties. Site and soil type was important in the survival of Grimm's alfalfa. Alfalfa will not grow in wet, poorly drained soil, so Grimm knew what he was doing when he managed his alfalfa on upland sites.

Gradually, news of Grimm seed spread, and a supply became available for sale. Seed companies carried all that they could obtain. Lyman, who had discovered it, rented land in the Dakotas, Idaho, and Montana, raised the seed there, and then sold the product under the brand Lyman's Grimm Alfalfa Seed.

By 1910 raising alfalfa was a significant part of farming for Carver County. The area became known as the Dairy Belt, with Grimm's place, where his son still farmed, as the "buckle." Signs in Carver County proclaimed, "Carver County, the Home of Grimm Alfalfa." Alfalfa's popularity grew, with the plant expanding to other northern states and becoming Minnesota's standard hay crop. During that decade Grimm alfalfa became the preferred variety in the North Central region of the United States. By 1930 Minnesota had 702,578 acres planted in alfalfa, compared with only 658 in 1900.

To honor Wendelin Grimm, in 1924 the Grimm Alfalfa Growers' Association erected a bronze monument at his original homestead property. Today the USDA credits Grimm alfalfa as being the source of most modern varieties grown in the northern United States. Alfalfa is planted on more than 18 million acres and is valued at $10 billion annually. An article on the "Wendelin Grimm Homestead" quoted Lawrence "Laddie" Elling, a retired professor from the University of

WHAT MINNESOTANS HAD TO SAY ABOUT ALFALFA

Alfalfa/forages (all hay) are Minnesota's third most valuable crop, and Minnesota ranks fifth in nationwide production. It is one of the most sustainable crops in terms of land stewardship and provides enough nitrogen for a subsequent corn crop when used in rotation. Alfalfa is an important component in beef, dairy, sheep, and equine diets.

CHELSEA R.

Grimm farmhouse, with memorial plaque. Three Rivers Park District

Minnesota Department of Agronomy, asserting that Grimm alfalfa was the most important agricultural crop developed in North America until hybrid corn in the 1930s.

The farm in Carver County is now maintained by the Three Rivers Park District and is open to the public at specific times of the year. More information is available at http://www.threeriversparks.org/parks/carver-park/grimm-historical-farm.aspx.

GRIMM'S HISTORIC FARMSTEAD

WENDELIN GRIMM built a farmhouse of Chaska brick about thirty miles west of Minneapolis, near the current town of Victoria. This was the Grimm home through the family's first years in Minnesota. In the 1870s, Wendelin's oldest son, Frank, and his bride, Rosella Pograbo, took over the original farm, and Wendelin and his wife, Julianna, moved to the new one between Victoria and Chaska.

By the mid-twentieth century, the farm was abandoned, trees were growing in the pastures, and the house was falling into disrepair. The Hennepin County Park Reserve District, now called the Three Rivers Park District, acquired the Grimm site in 1962. Twelve years later the farmstead was listed on the National Register of Historic Places. By 1993, the Minnesota Historical Society had become aware that many of Minnesota's historic agricultural sites were being lost and so had flagged the property as a high priority for preservation and restoration.

The restoration, begun in 1998–99, was completed three years later and celebrated with an open house on October 6, 2001. Now known as the Grimm Farm Historic Site, the farm is used to educate students and the public about the Grimms and early farming techniques.

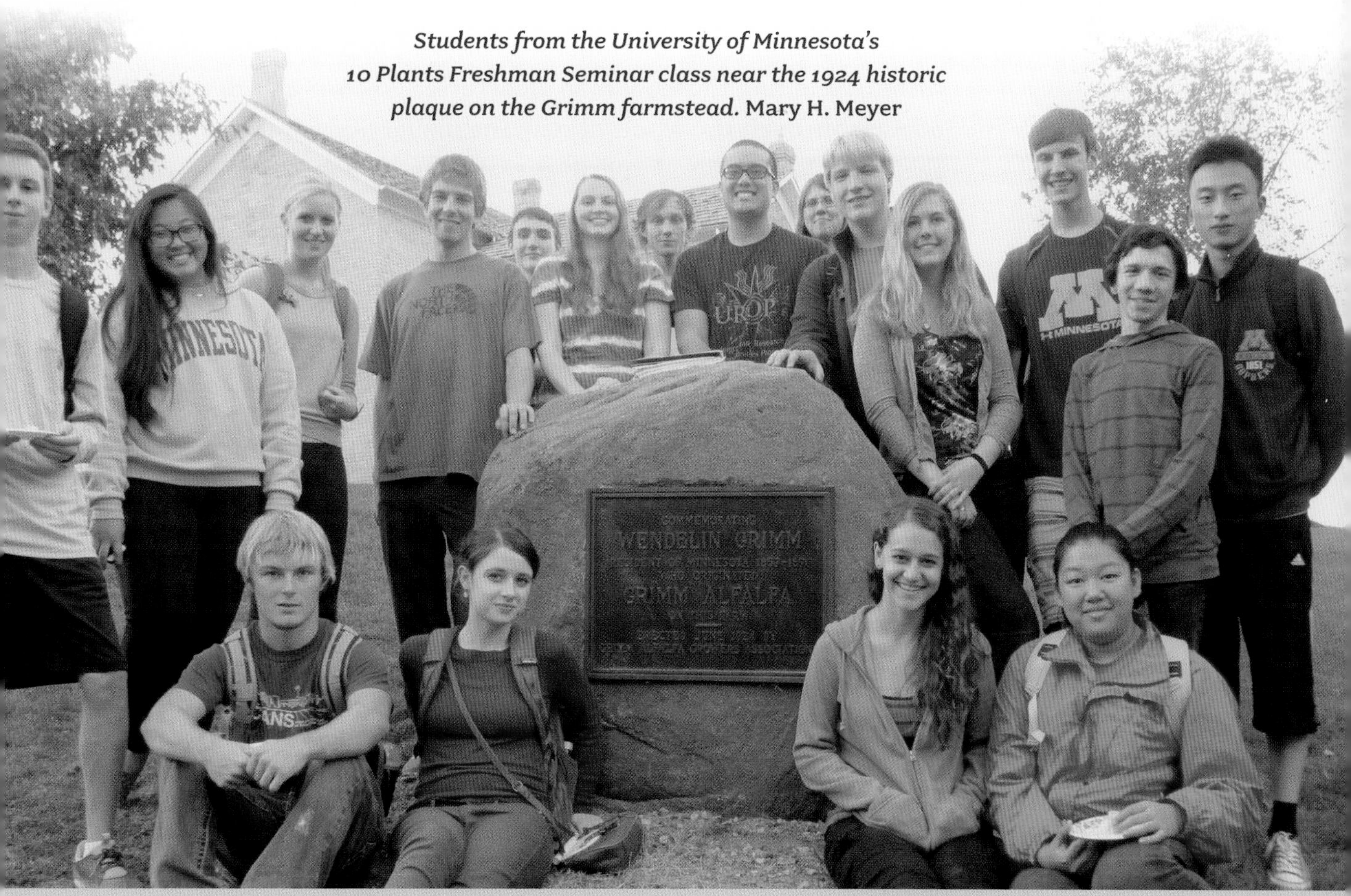

Students from the University of Minnesota's 10 Plants Freshman Seminar class near the 1924 historic plaque on the Grimm farmstead. Mary H. Meyer

PROMOTION EFFORTS

Early in the twentieth century a number of civic and agricultural organizations held meetings and festivals to encourage farmers to grow alfalfa. The first such event was the Corn and Alfalfa Exposition, held in Morris on December 8 through 12, 1913. A spectacular affair, it was sponsored by the West Central Minnesota Development Association to help improve living conditions in western Minnesota. More than 20,000 people attended, many arriving by a special train from the Twin Cities that stopped in small towns along the route. More than 200 Minneapolis businessmen were there, along with farmers and their families.

According to the *Willmar Tribune* of October 29, 1913, "The exposition will present heavy-weight speakers of national reputation, including the Honorable James J. Hill, and the president of the University of Minnesota, Dr. George E. Vincent." Workshops on growing corn and alfalfa were "interspersed with various entertainment features," including the Bookman Quartette, "known and deservedly popular through the whole state." The university's drama club presented a play, *A Pair of Spectacles*, and Mr. Cornie Wollan of Glenwood appeared with "his splendid whistling solos." Bands and orchestras added variety throughout the exposition.

The grand symbol of the event was a gigantic arch of alfalfa stretched across Atlantic Avenue in downtown Morris. The event was capped with a banquet on Friday night.

The following year Benson hosted the exposition, held November 26 through 28, 1914, with county agent Alfred Carlsted, local Benson civic groups, and the West Central Minnesota Development Association as the planners. Not to be outdone by Morris, Benson arranged for hundreds of corn and alfalfa exhibitors from sixteen counties, two great tents ("large as any ever carried by the Ringling Brothers," according to a *Willmar Tribune* article from November 18, 1914) complete with circus thrills, and "a dozen of the best bands in the Northwest." The streets were decorated with large arches and pylons, covered with corn, alfalfa, pumpkins, and strings of lights. The

WHAT MINNESOTANS HAD TO SAY ABOUT ALFALFA

This plant became the primary forage plant on farms across Minnesota.

DOUG T.

Alfalfa arch in Morris, Minnesota, created to celebrate the Corn and Alfalfa Exposition of December 1913. Courtesy of the Stevens County Historical Society, Morris, MN. All rights reserved

Queen of the Exposition, the winner of a competition that included girls from the sixteen participating counties, was crowned on the final day. Song fests, motion pictures, a husking bee, and a play were among the festivities. Three trains a day supplemented the regular rail service to accommodate the large number of visitors.

Amid all the fun, noted speakers presented serious workshops about farming and farm life: the honorable James J. Hill and President Vincent again but also Elbert Hubbard, philosopher, lecturer, and editor of *Fra* and *The Philistine*; W. A. Wheeler, alfalfa expert from South Dakota; and Dr. John W. Powell, religious director of the University of Minnesota. Outdoor presentations demonstrated methods for eradicating smut, a fungal disease, and testing seed corn. Hubbard, "the great American philosopher" according to the *Tribune*, delivered an address entitled "Getting Together."

The third year Ortonville was the host for the exposition. The September 28, 1915, issue of the *Willmar Tribune* described the site as being on the "shores of the beautiful Big Stone Lake. A spot known the country over as a summer resort." This time the town held the event one month earlier than previously, on October 13 through 15, in order to "eclipse in number and quality of exhibits, in educational facilities and in variety of entertainment all expositions of the kind ever held." Ortonville outdid itself with community sing-alongs, vaudeville acts, and excursions on the lake, as well as the usual lectures and workshops by well-known agricultural and civic leaders. A special train left the town of Willmar at seven o'clock in the morning, returning at ten thirty at night to transport visitors to and from the exposition. The fare was $4.36.

There seem to have been no more expositions of this kind in the next few years. Perhaps the buildup to World War I and the war's aftermath intervened. But Alfalfa Days resumed in the 1920s in a different part of Minnesota—the northwest. As one commentator noted in 1925 in the *Banker-Farmer*, "It was not until 1922 that the territory [Pennington County in northwest Minnesota] awoke to its opportunity in alfalfa raising." Here the bankers and businessmen of the county distributed a carload of registered Grimm alfalfa to local farmers to spur planting.

By 1924 the county was holding annual Alfalfa Days in Thief River Falls, with song and slogan contests for children, educational presentations, farm tours, entertainment, and decorations. Spanning the town's main street that year were a massive arch and a palace made of alfalfa bales, illuminated with electric lights. The gala celebrated 6,000 acres of alfalfa planted. Participants shouted the winning slogan, "Alfalfa's Nest, Minnesota's Northwest." The walls of the Thief River Falls auditorium resounded to more than a thousand voices singing "The Alfalfa Chorus," a song written by local schoolchildren and set to the tune of "The Battle Hymn of the Republic."

We have seen the green alfalfa
As it lay beneath the sun,
We have seen it dried and gathered
When the harvest time was done,
But through all the state and nation
We've not seen a richer ton
Than in our own Northwest.

Glory, glory to alfalfa!
Glory to our Northwest
treasure,
Glory, glory to alfalfa!
Its truth is marching on.

A second alfalfa song seems to have been adopted by Pennington County as well, this one sung to the tune of the folk song "Solomon Levi." In it, the farmer Peder Engelstad was recognized as the first in the county to plant alfalfa on his property (in 1905). Seeing the value to his cattle, he had been a big promoter of the crop.

We hail the home of Alfalfa
In Pennington County
It is the best of all the crops
That ever there could be
We live in the Alfalfa Zone
Of Minnesota state
And every farmer ought to be
A grower up-to-date.
(Chorus)
And there is Peder Engelstad
Of Pennington County
The Pioneer of Alfalfa
Of him we must take heed
He raises Guernsey cattle
And feeds Alfalfa Hay
Which makes the cream checks grow and swell
And thereby saves the day.
(Chorus)
Oh! Pennington County
Tra la la la la la la
Home of Alfalfa
Tra la la la la la la la la la la la la la la la la
We hail the home of Alfalfa
In Pennington County
It is the best of all the crops
That ever there could be.

Alfalfa promotions have continued into the present, but perhaps not with the same level of drama and entertainment as in these earlier years. Today they often have names such as Alfalfa and Forage Expo or Hay and Forage Expo. The 2015 expo, held in Cannon Falls, was billed as "the nation's largest hay and forage event" and promised to showcase the latest techniques, tools, and suppliers, with demonstrations on 200 acres of prime alfalfa and forages.

ALFALFA CULTIVATION

Alfalfa is an excellent crop for planting in rotation with corn. Nodules on its roots contain bacteria that take nitrogen gas from the air and convert it to nitrogen that plants can use. By this process, alfalfa replaces nitrogen in the soil which the corn has removed. Plus, it offers far better carbon sequestration (the long-term storage of carbon) than soybeans. It improves soil structure, adding valuable organic matter to the soil. Alfalfa has a high yield because it can

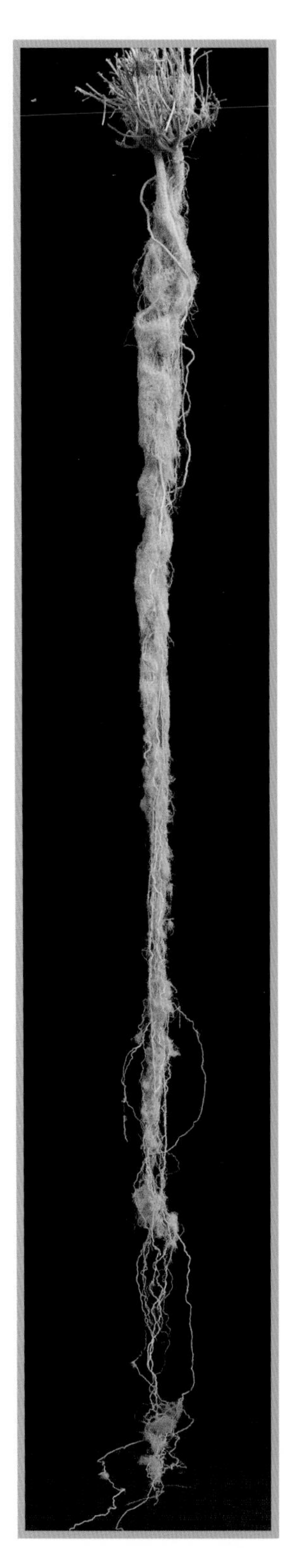

Alfalfa's long root system. David L. Hansen, University of Minnesota

ALFRED CARLSTED

ALFRED CARLSTED (1889–1972), the seventh of eight children, was born to Swedish immigrants on the farm he managed for many years, just south of Dassel, Minnesota. In 1910 he graduated from the School of Agriculture, the two-year technical high school operated by the University of Minnesota. Along with his brother Martin, also a School of Agriculture graduate, he operated the Carlsted Brothers Prospect Farm, where "a high state of cultivation and the most modern methods in every branch are used," according to the *Dassel Anchor* newspaper of 1911.

Carlsted became one of the state's first two University of Minnesota Extension county agents in October 1912, working in Swift County. He helped to organize the Farm Bureau and promoted changes such as diversified farming and progressive agriculture. He knew that the rotation of wheat on wheat was not sustainable. In "Why Alfalfa on Every Farm," a talk he gave in 1913, he described his own experience with growing alfalfa: "Alfalfa provides more feed per acre at less cost than any crop that we can grow. . . . [D]uring the five years we have never harvested less than three crops or four tons of good hay per acre, while some of it has yielded as much as six tons of good hay per acre."

Crop rotation was new, and change came slowly to farmers. Alfalfa's reputation for not being cold hardy was well known, and trying to grow Grimm's alfalfa was a risk. Arranging for Swift County farmers to pay twelve dollars per bushel for Grimm alfalfa seed (at cost) instead of the going rate of twenty to sixty dollars gave incentive to trying to grow alfalfa.

Although the early position of county agent was eliminated for political reasons, Carlsted continued to farm and teach others. He assisted in organizing two cooperative creameries and also Farmers, Boys, and Girls Clubs (organizations that provided fellowship and information for adults and youth); in addition, he promoted Minnesota 13, a much more productive type of corn for northern climates. No doubt in 1924 Carlsted attended the historic commemoration at the Grimm farmstead that recognized the importance of alfalfa selection in Minnesota.

Alfred Carlsted (left) examines alfalfa roots with a local farmer in 1913. Courtesy John Carlsted

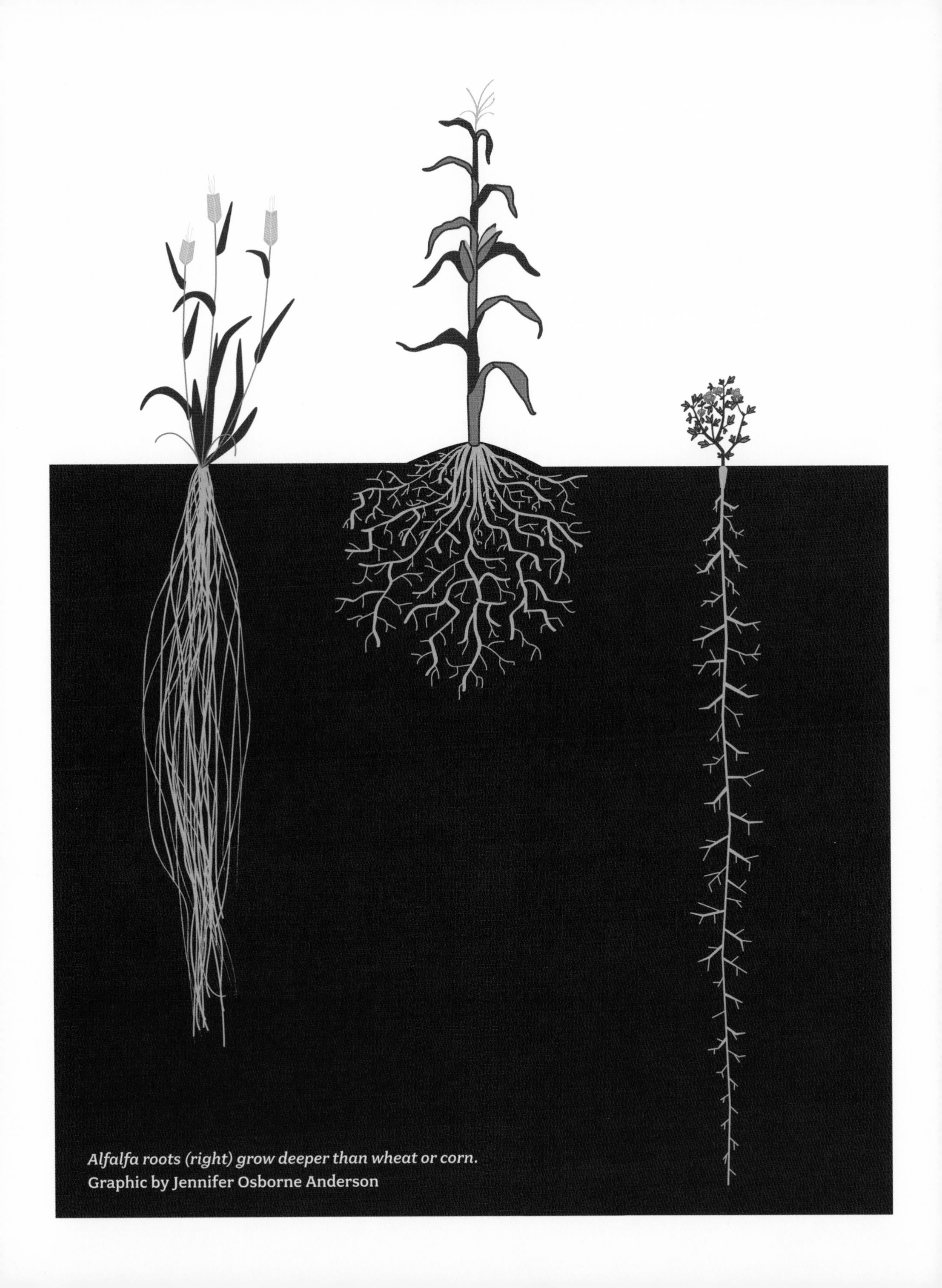

Alfalfa roots (right) grow deeper than wheat or corn.
Graphic by Jennifer Osborne Anderson

DID YOU KNOW?

 Alfalfa has been used as an **HERBAL MEDICINE** for more than 1,500 years.

 The **SCIENTIFIC NAME** for alfalfa is *Medicago sativa.*

 Alfalfa is native to **ASIA MINOR** and the **CAUCASUS MOUNTAINS**.

 Alfalfa was grown by the **PERSIANS, GREEKS,** and **ROMANS**.

 The Arabians called alfalfa "father of all foods" after observing that it made their **CATTLE STRONGER**.

 Alfalfa was first introduced into North and South America **IN THE 1500S** by the Spanish conquistadors, who used it to feed their horses.

 Alfalfa is known in many other parts of the world as **LUCERNE**, **MEDIC**, **CHILEAN CLOVER**, and **BUFFALO GRASS**.

 There are **74 MILLION ACRES** of cultivated alfalfa in the world, about 18 million of which are in the United States.

 Alfalfa is a **LEGUME** that belongs to the pea and bean family.

 The alfalfa plant has a **DEEP ROOT SYSTEM**, sometimes stretching more than forty-nine feet.

 Alfalfa is used in **ORGANIC GARDENING** as a fertilizer, since it adds value to the soil through its nitrogen-fixing ability.

Alfalfa flowers' blue color is thought to be the reason for the early name lucerne, *similar to lapis lazuli, the ultramarine blue mineral.* Patrick J. Alexander, USDA images

2
AMERICAN ELM

We loved the trees. We loved the canopy, the way they arched over the street—it was so beautiful.

ST. PAUL RESIDENT, 2014

Thirty years ago, American elms, with their vaselike form, arched gracefully over the streets of nearly every city and town in Minnesota. The green canopy created shade and cooled the road, while providing homes for birds and food for butterflies and moths. The high, leafy cathedrals added elegance. Elms were favored for their fast growth, longevity, and adaptability to many sites. They grew well in the urban environment, with its air pollution, road salt, and compacted soil. Indeed, many individuals and towns considered the American elm "the perfect tree." Native throughout the state and tolerant of our hot summers and cold winters (to forty-four degrees below zero), elms were planted in backyards and on boulevards.

The elm did not grace Minnesota streets alone. It had been the favorite street tree of communities across America from the early days of town-making. Set out in long rows, these plantings made a

Lombard Street in St. Paul before Dutch elm disease hit. MNHS collections

SNOW
EMERGENCY
ROUTE

splendid architectural picture. Native to North America, the American elm occurs from Nova Scotia west to Alberta and Montana, and south to Florida and central Texas. A survey from 1937 reported that more than 25 million American elms grew in cities and towns across the nation. Sacramento had as many elms as New Haven, Connecticut; Dubuque, Iowa, had more than Springfield, Massachusetts. Places like Elmhurst, Illinois; Elm Grove, Wisconsin; and Elm Point and Elm Creek, Minnesota, incorporated the tree into their names. Innumerable "Elm Streets" were created as New Englanders and other migrants moved westward. Indeed, Elm Street is thought to be the most common street name in America. In Minnesota there is an Elm Street in towns from Alexandria to Waconia, from Albert Lea to White Bear Lake. The elm is the state tree of both Massachusetts and far-flung North Dakota. The elm has lived up to its name: *Ulmus americana*, American elm.

As early as the 1740s, the Swedish botanist and explorer Peter Kalm, after coming to America, noted, "The American Elm grows in abundance in the forests herewith." A bit later the French botanist André Michaux charmingly described the American elm as "the most magnificent vegetable in the temperate zone." And in 1780 Samuel Vaughn, who designed State House Square in Philadelphia, now named Independence Square, planted elms in a double allée along the north-south path. There our Founding Fathers would have walked day after day. In Washington, DC, too, elms were always a significant part of the landscape, though many were cut down during the Civil War, when most trees were felled to build the forts that encircled and defended the capital.

WHAT MINNESOTANS HAD TO SAY ABOUT ELMS

The entire state was wounded by their sad departure.

WAYNE

Favored in parks, on college campuses, and on private lawns as well as along streets, elms were ubiquitous. Andrew Jackson Downing, a noted mid-nineteenth-century American landscape designer, was especially fond of American elms, whose broadly spreading branches angled outward. He used them for city tree plantings and for the estate landscapes he designed, stating in his *Treatise on the Theory and Practice of Landscape Gardening*, "Let us now claim for the elm the epithets graceful and elegant."

Frederick Law Olmsted, designer of New York's Central Park and often called the "father of American landscape architecture," used elms in the original Promenade of the park's mall. Here he planted a quadruple row of American elms that remain today, with their distinctive arches extending over the park's widest walkway, where they are one of the park's most photographed features. Taking care of the trees along this quarter-mile pathway is a full-time job for the Central Park Conservancy's tree crew. These mature elms form one of the largest and last remaining stands of American elms in the United States.

Throughout American history the elm has acquired honorable status. Colonists planted them to mark weddings and births. They became centerpieces of towns and villages. Under the elm, treaties were signed, revolutions declared, and oaths of office taken—and so we have the Justice Elm, the Divine Elm, the Washington Elm, the Liberty Tree, and the William Penn Treaty Elm.

In Minnesota, Northfield has claimed at least two such famous trees, the St. Olaf Elm on Forest Avenue and the Ladies Hall Elm on the St. Olaf College campus. The Ladies Hall Elm, growing by the women's dormitory, was noted for the enormous hollow in which many folks delighted to sit and be photographed. Due to age and disease, both large trees had to be removed in the 1920s. Lyle, Minnesota, in Mower County, had the Big Elm Tree, which stood on Thrond Kleppo Richardson's farm. Here the first Norwegian immigrants met in 1853 to organize

Thonny, Elsa, and Volborg Felland photographed by their father, O. J. Felland, in Ladies Hall Elm on the St. Olaf campus, May 23, 1895. Courtesy of the St. Olaf College Archives, Northfield, MN

the Six Mile Grove Lutheran congregation. Services were held under the tree and in nearby homes until the church was built. The elm is long gone, but in 2009 Six Mile Grove Lutheran celebrated 150 years as a congregation and remembered the tree on Richardson's farm.

As a tribute to those Hennepin County soldiers, sailors, and nurses killed in World War I, the Minneapolis Park and Recreation Board in 1921 planted the Victory Memorial Drive with 1,200 elms, in four straight lines, along each side of the drive. A plaque with the name of one of the deceased was placed beneath 568 of the trees. At the time, the Minneapolis park superintendent, Theodore Wirth, stated, "Straight lines of gigantic trees will represent in a

An allee of American elms.
iStock.com marekuliasz

Elmer the Elm visits schools in Minneapolis. Minneapolis Park and Recreation Board

most impressive manner both the strength of the combined army and the sacrifice of those who fell." The original trees were Moline elms, which were not winter hardy. These were later replaced with local American elms. The trees made an elegant memorial until most were felled by Dutch elm disease in the 1970s and 1980s. Hackberry trees, whose profile is similar to the elm, were planted to replace the diseased trees. In 2010 the Victory Memorial Drive was refurbished and a mixture of tree types planted.

At times the elm is treated playfully. The Minneapolis Park and Recreation Board has used Elmer the Elm Tree as the official mascot for its Forestry Department since 1976. "Elmer" has visited school kids and other groups to point out the value of trees, reminding them of the large elms that once lined Minneapolis streets.

As easterners moved to Minnesota in the 1850s through the 1870s, they brought their notions of enhancing city beauty. According

FAMOUS ELMS IN AMERICAN HISTORY

THE TREATY ELM, ALSO CALLED THE WILLIAM PENN TREATY ELM (PENNSYLVANIA). Legend holds that in 1682 in Philadelphia, William Penn, founder of Pennsylvania, reached a peace treaty with the Lenape Turtle clan under an elm tree growing in what is now Penn Treaty Park. The artist Benjamin West incorporated the tree and the event into a painting after hearing stories about the location for the agreement. Though no documentary evidence exists, one elm in the park was long thought to be the spot for Penn's peace treaty. In 1810 a storm blew down the enormous tree—then measured at twenty-four feet in circumference and estimated to be 280 years old.

BOSTON LIBERTY ELM (MASSACHUSETTS). In Boston in 1765 a large elm at the corner of Essex and Washington Streets became the rallying point for the colonists protesting the hated British Stamp Act. The tree, with its wide-spreading branches, was growing in an area known as the Neighborhood of Elms, in the city's original South End. Here the patriots met on August 14 and, in the first show of defiance against the crown, hung an effigy of Andrew Oliver, the man chosen by King George III to impose the Stamp Act. On September 10 of that year, the Sons of Liberty nailed a bronze plaque to the trunk of the tree, identifying it as the "Tree of Liberty."

The tree became a point of pride and rebellion for the patriots and a symbol of ridicule for the Loyalists. People of all classes flocked to the tree to post notices, hear speeches, and hold outdoor meetings. In September 1775, during the ten-month siege of Boston, British soldiers cut down the tree with much "grinning, sweating, swearing, and foaming with malice diabolical," according to the September 9, 1775, issue of the *Constitutional Gazette*. Today a small plaque marks the spot.

ST. OLAF ELM (NORTHFIELD, MINNESOTA). Probably old when Northfield was founded, the large elm on Forest Avenue began its official career as the St. Olaf Elm in 1897, when the city council passed an ordinance prescribing punishment by "fine or imprisonment on anyone who shall willfully cut, injure, or deface the said St. Olaf Elm." Previously the tree, which was featured on postcards and in St. Olaf yearbooks, was a much-admired landmark but was not officially named or protected.

Earlier, in about 1894 or 1895, the elm had been slated to be cut down, but tree lovers came to its rescue. A baseball game between St. Olaf and Carleton College was organized by Professor Andrew Fossum, with the proceeds from the admission fee to be used to surround the tree with an iron railing and provide for tree surgeries.

A St. Olaf student from the graduating class of 1918 composed the following poem to the revered tree:

> O, sturdy elm, in thy majestic mien,
> In thy rich shades, thy glory, we delight.
> Thou hast no peer, nor is thine equal seen
> Amongst thy forest mates. For in thy might
> And strength thou hast o'ertopped them, every one.
> Endured the tempests of many a stormy day,
> And still, in pride thou hold'st aloft the crown.
> That marks thee king of all but Time's decay.

Unfortunately the tree eventually did succumb to time's decay and had to be cut down in 1920.

to one of the first historians of Minneapolis, "The original New England settlers platted the streets to reflect order and prosperity, with boulevards lined with oak and elm trees." Early photographs of streets on Nicollet Island, homesteads in St. Cloud, and farms in the country often depict folks standing proudly beside the elm saplings lining the road or ornamenting the house. The elm was so easy to transplant that the newly arrived immigrants could simply pull a small tree from a swamp, plant it into their plot, and expect it to take hold and thrive.

St. Paul house with sapling elm in front boulevard, 1900. MNHS collections

USES FOR ELM TREES

BESIDES BEING ADMIRED for their beauty and stateliness, members of the species have been useful trees for centuries. The earliest written reference to the elm occurs in a list of military equipment at Knossos, the major city of Crete, from the Mycenaean period, around 1400 BCE. The document lists elm as the material used for chariots and wheels. The ancient Greeks made plows out of elm wood, and the Romans, too, used it for chariot construction.

The interlocking grain, which resists splitting, made the elm useful for wagon wheel hubs, chair seats, and coffins. The long, straight trunks were prized as a source of timber for keels in ship construction; in the Middle Ages, bow makers valued the elm for the longbow. The Iroquois used elm bark to make many items, including ropes, canoes, and utensils.

DUTCH ELM HITS

Unfortunately, the elm's popularity led to a monoculture in many towns across America, creating the perfect conditions for disease to spread. Dutch elm disease (DED) entered the United States in 1931 on a shipment of logs from France. Within four to five years, scientists could trace its movement through the country along rail routes, and by 1970 it had killed an estimated 77 million trees in America. The disease, which was first diagnosed in Minnesota in 1961, is caused by the deadly fungus *Ophiostoma ulmi* and most recently *O. novo-ulmi*. The pathogen can be transmitted in two ways. Elm bark beetles, both native and European, can carry spores of the fungus and spread the pathogen to healthy trees when feeding on branches and branch crotches. Once a tree is infected, the fungus moves through the xylem (the tree's transport tissue), causing a disruption in water flow, which results in the characteristic wilt symptoms. A second way that the pathogen can spread to healthy trees is through root grafts. When trees are planted close together, as can be seen in a boulevard planting, the roots of the trees can become grafted and allow for the movement of the pathogen from an infected tree to a neighboring healthy tree.

And though the zenith of the crisis is behind us, DED is still a serious problem for any remaining trees. It is estimated that only one in 100,000 American elms is DED-resistant. Before the disease hit, Minnesota had more than 140 million elms. As it peaked here in the 1980s, trees wilted and weakened in waves, leaving few elms in the urban environment. Elms still survive in woodlands, as well as in suburbs and cities. Most of these are small to medium-sized trees that are relatively isolated from other elms and thus not exposed to the pathogen.

Work on two fronts has made a significant contribution to restoring elms to the landscape. Individuals and institutions can protect the remaining trees with fungicidal injections at the base of the tree. Though the treatments don't guarantee survival, they can provide a tree with up to two years of protection from DED. In addition

DID YOU KNOW?

- The **SCIENTIFIC NAME** for American elm is *Ulmus americana*.
- Other names for the tree are **SOFT ELM**, **WATER ELM**, and **WHITE ELM**.
- The **IROQUOIS** used elm bark to make canoes, ropes, and utensils.
- A strong wood, elm was also used for **SHIP DECKS** and **WAGON WHEEL HUBS**.
- The wood is prized by **PIANO AND FURNITURE MAKERS** for its strength, weight, and toughness and is commonly used in making **HOCKEY STICKS** because it can bend without splitting.
- Under ideal conditions the American elm can live for **300 YEARS**.
- American elms can reach **MORE THAN 100 FEET** in height and four feet in diameter.
- **GEORGE WASHINGTON** is said to have taken command of the Continental Army under the **WASHINGTON ELM** in Cambridge, Massachusetts, on July 3, 1775.
- Elm trees first appeared in the **MIOCENE EPOCH**, about 20 million years ago.
- **DUTCH ELM DISEASE** is so named because it was discovered by scientists in Holland in 1917.

3

APPLES

Never move to Minnesota—
you can't grow apples there!

HORACE GREELEY, EDITOR, *NEW YORK TRIBUNE*, 1860

Apples and autumn: the two go together. Turning leaves, cool evenings, apples at roadside stands—all are sure signs of fall. For many Minnesotans, heading to the orchard to harvest apples has become an autumn tradition; savoring apple pie and hot apple cider seems even more enjoyable when days are crisp. Fortunately, though, today Minnesota apples can be enjoyed year round! Here in Minnesota, we are lucky to have a large selection of apple varieties that can be grown around the state, most of them developed locally at the University of Minnesota.

Initially the climate made Minnesota a difficult place to grow apples. Our winters are notoriously bitter, the growing season is short, and rainfall is variable. Early Minnesotans were frequently taunted about growing conditions in the state.

"What can be raised away up there in Siberia?" their East Coast friends asked. In rejoinder the *Minnesota Republican* in 1855 ran a series of articles discussing the fine corn, good soil, and large turnips. Civic leaders understood that successful horticulture was essential to a prosperous state. But growing vegetables was one matter; raising fruit was quite another.

The "king of fruit," the apple, was what Yankees and immigrants wanted. It is difficult for twenty-first-century Minnesotans to understand the need for apples. To us, the apple is a sweet snack or a crucial ingredient of a pie. To the earliest Minnesotans it was a versatile and essential food. Besides wanting a fruit they could eat out of hand, they wanted one that would keep through the winter when food was scarce. Apples could be dried and made into pies and also could be used to make vinegar, important in preserving meats and vegetables. Minnesotans fermented the fruit to make hard cider, cheaper than beer and a favored drink when early farmers and their families feared water might be contaminated.

Apples, especially the famous Honeycrisp, were the public's most nominated plant for the 10 Plants project. David L. Hansen, University of Minnesota

LEFT: John Harris of La Crescent, about 1895. MNHS collections

BELOW: John Harris's orchard. MNHS collections

Early diaries and newspaper stories record newly arrived Euro-Americans' attempts to grow apples from back East. Their efforts were rarely successful. This account in *The History of Freeborn County* is typical:

> The honor of establishing the first good-sized orchard in the county belongs to the Rev. Isaac W. McReynolds . . . who planted apple seed in 1858. These trees in due time came on and bore considerable fruit. . . . However, like most of the early seedling orchards that were grown from promiscuous seed gathered from eastern orchards, they carried with them in their ancestry no special adaption to the climate of the West, and one by one they succumbed to severe winters and droughty summers, till at the end of twenty years very little was left to show for the effort that had been put forth.

What was needed were apples with "special adaption to the climate of the West." The two men most responsible for Minnesota's early successful apple crops were John S. Harris of La Crescent and the colorful and eccentric Peter Miller Gideon, who set up his home near Lake Minnetonka. For a time the problems associated with growing apples seemed insurmountable. Harris, who has been called the "father of orchards in Minnesota," reported the reaction of other farmers at the 1866 state fair when he and Colonel D. A. Roberts of St. Paul proposed a fruit growers' society. "We were looked upon by many," said Harris, "with serious suspicions of our sanity, but our hopes and enthusiasm became contagious and has extended to every part of the state." His exhibit of twenty apples gave the group a boost. The Minnesota Fruit Society, organized that year, later became the Minnesota State Horticultural Society, which endures to the present.

Harris planted his first apple trees in La Crescent in 1857 and experimented with them until he was able to grow trees hardy enough to withstand the severe Minnesota winters. He planted thousands of apple trees and hundreds of varieties, a full half of which he said were complete failures. His efforts attracted other orchardists, and the town became known as the Apple Capital of Minnesota. Since

WHAT MINNESOTANS HAD TO SAY ABOUT APPLES

Apples developed by the U of M create variety in our diets, our yards, and excitement in our gardening community, and very favorable attention to our beloved U of M! Our commercial growers have reaped billions of dollars in benefits, including those farmers who sell at the roadside stands. Who doesn't love Minnesota apples?

CYNTHIA S.

La Crescent is noted for its apple orchards and early apple history. Courtesy La Crosse Sign Company

1947 La Crescent has celebrated this apple heritage annually with a weekend festival known as Applefest.

Like Harris, Peter Gideon had a determined spirit. In 1853, shortly after arriving from Ohio with his wife, Wealthy, and their children, he set up shop on Lake Minnetonka, homesteading on the side of the bay that now carries his name. He began experimental plantings of numerous named varieties of apples and other fruit trees. By the tenth year every tree except one seedling Siberian crab apple had fallen victim to the severe Minnesota winters.

With only one cow and fewer than twenty chickens to his name, and a wife and children to feed, Gideon was desperate. But, being

determined, Gideon sent his last eight dollars to Bangor, Maine, for seeds and scion wood (grafting material) and waited for spring in a homemade jacket he had sewn from two old vests. By his own account, it was a winter suit "more odd than ornamental."

Gideon's persistence was rewarded. On receiving from Maine seeds of the Duchess, Blue Pearmain, and Cherry crab, he crossed the larger varieties with the hardy crab apple. The result, in 1868, was the Wealthy, named in honor of his wife—a good-looking, tasty fruit that could withstand the region's climate.

Peter Miller Gideon, about 1890. MNHS collections

The Wealthy—which is full-sized, bears regularly, and has reasonable keeping qualities—became immensely popular. By the early twentieth century, it was one of the top five apples grown nationally, and it is still grown today. In addition, the success of Wealthy fueled the quest for more Minnesota apples of excellent quality, giving other growers the incentive to continue the work Gideon had started.

WHAT MINNESOTANS HAD TO SAY ABOUT APPLES

Apple production has become a major economic and extracurricular activity.

LUKE S.

The success of the Wealthy apple led to the Minnesota legislature's approval of state-supported experimentation in fruit breeding. In March 1878, at the urging of the Minnesota State Horticultural Society, the legislature established the Minnetonka Fruit Farm on a 116-acre plot next to Gideon's land. Gideon himself was appointed superintendent, and the farm was placed under the jurisdiction of the University of Minnesota.

For twelve years Gideon supervised the fruit farm, continuing apple experimentation, particularly the development of "long keepers," but introducing no new varieties during that time. When he retired in 1889, the land was sold. Gideon's eccentric nature and his determination to operate as a lone wolf probably contributed to the farm's demise, but his work there encouraged later botanical developments. For example, in 1883, the Minnesota legislature established a series of experiment stations for testing ornamentals and vegetables as well as fruit. The results of their experiments were published twice a year in the *Minnesota Horticulturist,* the state's primary horticulture journal, known today as the *Northern Gardener* magazine.

In 1907 the State Horticultural Society lobbied the legislature for a fruit breeding and testing farm to be a part of the University of Minnesota's Department of Horticulture. In its annual report to the legislature, the Horticultural Society said, "Such a farm is greatly needed to place the state's horticultural work in a satisfactory condition." Further, the report stated that the $16,000 gained from selling the Minnetonka Fruit Farm could be used to purchase new acreage. This request was approved by the legislature the same year, and a seventy-eight-acre tract five miles west of Excelsior was purchased in 1908 as the location of the Fruit Breeding Farm. The spot was chosen because it could be easily reached by train both from Minneapolis and from the university's St. Paul campus. Additional

The Wealthy apple, Gideon's first success. Lake City Nursery catalog, Andersen Horticultural Library, University of Minnesota

land was purchased in 1920 and 1921, bringing the total to approximately 230 acres.

The Fruit Breeding Farm was incorporated as part of the university's Department of Horticulture, with Charles Haralson named its first superintendent. In taking the job, Haralson sounded a note that has been echoed through the decades: "We are near the northern limits of fruit growing and therefore must grow seedlings from our hardiest varieties to secure the best results." Haralson held the position of superintendent until 1923, and his longevity at the farm contributed to its successes.

The winter of 1917–18 was a "test" winter—unusually severe—and apples died at a greater rate than usual. But there were survivors, most notably the progeny of Malinda, originally from New England. These survivors led to successful apple releases in the 1920s, the Haralson and Beacon, which became the basis for Minnesota apple breeding up through the present time. DNA testing has shown that Wealthy, Gideon's introduction, is the likely parent of the still popular Haralson.

In 1967 the Fruit Breeding Farm was renamed the Horticultural Research Center (HRC), reflecting its expanded mission to include ornamentals as well as fruit. And in 1987 the HRC merged administratively with the Minnesota Landscape Arboretum, acknowledging the institutions' long cooperative relationship.

TODAY'S PROGRAMS

Today, the program at the university, one of only three university apple breeding programs in the country and the last major fruit breeding program in the Midwest, has become known for producing hardy fruit varieties. To date, the university has introduced twenty-eight apple varieties, including Fireside in 1943, Sweet Sixteen in 1978, and the very popular Honeycrisp in 1991. Newer varieties include SnowSweet® in 2006 (a wilding cultivar) and Minneiska, whose fruits are sold under the SweeTango® brand, both introduced in 2006. In 2008 Frostbite™ (bred in 1936) became commercially available.

Fruit Breeding Farm of the University of Minnesota, early 1900s. Andersen Horticultural Library, University of Minnesota

Because of this extensive research and development, orchards not only in the Midwest but throughout the world have a broad range of apples to plant. Fruit varieties developed here, including Regent, Prairie Spy, Beacon, and Red Baron, now represent three-fourths of the state's fruit production.

Fruit breeders are patient people—they have to be. It takes decades to discover and bring to market a new plant. In fact, for apples the process usually takes about thirty years. Twenty thousand seedling trees are growing at the HRC at any one time; of these, only one is likely to be named and released to the market. Fortunately, every now and then, a fantastic fruit will come along that makes the wait worthwhile.

MINNESOTA'S DISTINCTIVE APPLES

MINNESOTANS are fortunate to have one of the most distinctive assortments of apples in the country. Some unique varieties found here are Fireside, State Fair, Beacon, Prairie Spy, Haralson, Red Baron, Keepsake, Sweet Sixteen, Frostbite™, Centennial, Viking, and Malinda.

The Haralson has been a front-runner for years. Its tart flavor and firm texture make it perfect for pies, good for juices, and tasty for eating "out of hand." Fireside, a longtime favorite, was introduced in 1943 and is sometimes called the Minnesota Delicious. It is a big apple that ripens in October with a sweet flavor and firm texture. Keepsake, though not a beautiful apple, is hard and crisp, with a flavor akin to sugarcane. True to its name, it is a great keeper and will last for months in cool, moist conditions. SweeTango® fruits ripen in late August and early September and are known for their juicy, crunchy sweet-tart taste.

Much of the credit for the wealth of apple varieties in the state must go to the University of Minnesota's Horticultural Research Center and its extensive apple breeding program, which has been developing fruit for more than a hundred years. In addition, many local orchardists have found success by devoting land to these new varieties that will thrive in the northland.

For scientist David Bedford, the HRC's chief apple breeder, that fruit was Honeycrisp. When Bedford arrived at the center in 1979, among the many apples he evaluated was MN1711 (later named Honeycrisp)—a variety that had been growing for twenty years and had already been tagged "discard" by previous evaluators. Bedford noted, however, that the original MN1711 seedling was growing in the worst site, one that was low and damp—conditions that had resulted in 80 percent of the trees in that area dying during a difficult winter. Bedford decided to remove the discard tag and give the tree another chance.

In the 1980s Bedford and Dr. James Luby, professor of fruit research, tasted MN1711 apples again. During tasting season, scientists sample hundreds of apples a day (it's not unusual for them to sample 500 apples in the peak of the season), but the two knew immediately they had something special. MN1711 was attractive and sweet-tart, and it had an explosively crisp texture described by one grower as a "one-of-a-kind crispness that no previous apple had ever possessed." This crispness is present because the fruit has much larger cells than most apples. They burst when bitten, filling the mouth with juice. In 1988 Bedford and Luby decided to send this one to the public, and in 1991 Honeycrisp was given its name and released to growers. The apple soon took the world by storm, becoming popular in Minnesota, across the United States and Canada, and eventually around the world.

With more than 15 million trees planted worldwide, the Honeycrisp is the university's most successful introduction to date. Apples are propagated vegetatively (usually through budding, a special type of grafting) to maintain their genetic integrity. All the Honeycrisp

David Bedford, "discoverer" of the Honeycrisp apple. David L. Hansen, University of Minnesota

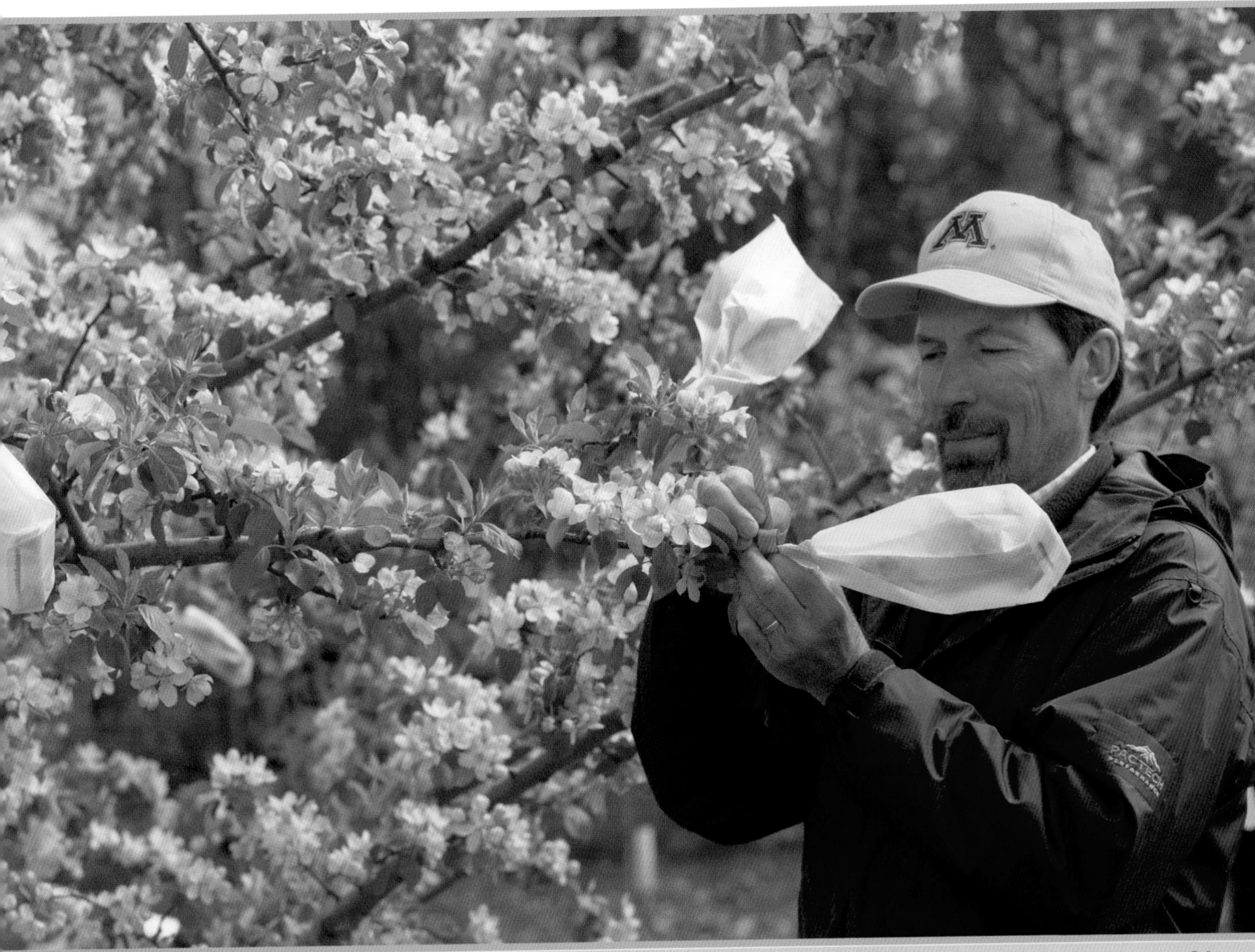

apples in the world have come from the original plant, discovered in Minnesota.

Known in Europe as Honeycrunch®, the apple is good for home gardens or commercial operations. It sells well because it is very crisp, has a balance of sweetness and acidity, and has excellent storage qualities. It helped renew Minnesota's apple-growing industry and brought much-needed revenue to small and medium-sized orchards.

In 2006 the development of Honeycrisp was selected as one of the top twenty-five innovations of the decade by the Better World Report. Prepared by the Association of Technology Managers, the report recognized academic research and technology applications that have improved our way of life. In granting the award, the report noted that Honeycrisp, with "its almost magical properties," helped revive a shrinking apple industry. In Minnesota more than 40 million Honeycrisp fruits are produced each year.

Apples are a $14 million industry in Minnesota. In 2014 the state had commercial apple orchards in seventy-eight of its eighty-seven counties, producing more than 11 million pounds of fruit annually on approximately 4,000 acres. Farmers can expect 300 to 500 bushels (up to forty pounds each) per acre. There are more than 117 retail orchards throughout the state, selling a distinctive assortment of apple varieties.

THE DEVELOPMENT PROCESS

Developing new varieties is a long, demanding process. From the seedlings, grown from crosses that hopefully will produce superior fruit, breeders will take two buds and graft them onto dwarf rootstock. Using dwarf rootstock speeds up the process because these trees produce normal-sized fruit in four to five years, not the seven to ten years typical of a full-sized tree. Additionally, dwarf trees save valuable space in the orchard, a significant consideration when thousands of trees are grown.

Many seedlings will be culled as juveniles when they exhibit diseases, winter injury, or other problems. Less than one percent of the seedlings will be selected for further testing. Once saved, each plant is given a number, such as MN1523, a designation it will retain until it is discarded or eventually named and introduced as a new variety.

Once the tree begins producing fruit, two to five years after grafting, tasters will test the apples for flavor and texture and observe them for appearance. From the initial quantity, 2,000 to 3,000 trees are discarded in their first year because they do not produce fruit that is unique. Those that pass the initial testing are thoroughly evaluated by numerous breeders and taste panels for twenty-five different characteristics. Overall, about one in every 20,000 of these different apple seedlings tested is named and released to the public.

Apple orchard at peak of harvest. David L. Hansen, University of Minnesota

WHAT MINNESOTANS HAD TO SAY ABOUT APPLES

Locally grown apple trees promote lower transportation costs. The apples are beneficial for human health and trees look great in the landscape.

IWONA B.

In recent years, the process has been streamlined a bit with technical innovations. Using DNA markers, breeders can select traits for the apple before the tree is planted in the orchard. Disease resistance, fruit color, and fruit texture can be selected based on the DNA fingerprints of young seedlings, increasing the efficiency of the fruit breeding program.

The process of patenting and releasing the budded trees to growers, increasing production, and getting the fruit into the marketplace takes about fifteen to twenty years. The Honeycrisp apple, which has become well known in recent years, was actually released in 1991; the cross was made in the 1960s, making this variety technically about fifty years old. Now Honeycrisp is the number one apple (in terms of pounds of fruit) in Minnesota and the number six apple nationwide.

The apple was chosen as one of the 10 Plants because of its great influence on the state. Apples were the number one plant nominated by the public when nominations were called for in the spring of 2012. Nominators cited the health benefits of apples, their economic impact on the state, and the importance of apple research spearheaded by the University of Minnesota. Apples promote health, beauty, and a true sense of community within the state.

Minnesota apple experts predict that the fruit will remain a popular crop and retail item. Most Minnesota-grown apples are purchased locally at orchards, roadside farm stands, and farmers' markets. Today more consumers are interested in buying locally grown foods and value the chance to meet the farmer. Visiting nearby orchards to pick or select apples is often part of the fall experience, with many Minnesota options.

Researchers note that we are seeing a major shift in the apples that are available. Through the mid-twentieth century, chance seedling apples that were found along fencerows or in the woods and pasture were our most important varieties. These include the McIntosh, Red Delicious, Golden Delicious, Granny Smith, Braeburn, and others. The efforts of apple breeders came to market in the late twentieth century with Gala, Fuji, Pink Lady®, and Honeycrisp. The twenty-first century will see many new varieties bringing delight to apple eaters.

NUTRITION AND HEALTH BENEFITS OF APPLES

AN APPLE A DAY may not keep the doctor away, but it will certainly benefit your health. Research has found numerous links between apples and weight loss, brain health (including lower incidence of Alzheimer's disease), reduction in certain forms of cancer (breast, pancreatic, colon, prostate, and bowel), and lung and heart health.

Apples are a refreshing snack that is low in calories but high in dietary fiber, and they contain no fat, sodium, or cholesterol. A medium-sized apple with its skin has about eighty-one calories, four grams of fiber, and twenty-one grams of carbohydrates.

Apples contain an impressive array of antioxidants and are a good source of potassium and vitamin A, vitamin C, and the B complex vitamins. The antioxidant quercetin in apples can help boost the immune system.

Because apples contain pectin, a type of fiber, people who eat two apples a day may lower their cholesterol as much as 16 percent. Those who consume apples and apple products also are likely to have lower blood pressure and trimmer waistlines, resulting in a reduced risk of metabolic syndrome.

For a more complete list, see the US Apple Association website "Apple Health Benefits."

THE CHRONOLOGY OF GETTING AN APPLE TO MARKET

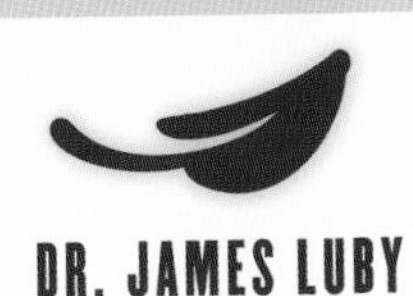

DR. JAMES LUBY

YEAR 1, SPRING: Pollen is collected from one parent, a cross is made by hand on flowers of the female parent. The fruit is allowed to mature, carrying the seeds of the cross.

YEAR 1, FALL: Apples from the controlled crosses are harvested and the seed collected.

YEAR 2: Seeds are planted in the greenhouse.

YEAR 4: The new trees are grafted onto hardy dwarfing rootstocks and field planted. Over the next few years they will be regularly inspected for disease and cold damage.

YEARS 6–20: The trees start bearing fruit and are tested extensively for hardiness, disease resistance, and fruit quality. More than 99 percent of the original seedlings will be eliminated.

YEARS 21–22: Initial field propagation is undertaken by nurseries to create enough stock to make available for sale.

YEAR 23: Substantial stock is available to apple growers and home gardeners.

YEARS 27–30: The first significant amount of fruit becomes available in markets.

APPLES *and the* ENVIRONMENT

THE US–BASED Environmental Working Group is a nonprofit research and advocacy group dedicated to informing people about issues of public health and the environment. The organization publishes an annual list of the "Dirty Dozen," fruits and vegetables that contain the most pesticide residue when sold. Apples always rank high on the list, with at least 98 percent of them testing positive for at least one pesticide residue.

Fortunately, residues for all pesticides on apples have been well below the Environmental Protection Agency's tolerance for fruit. It's helpful to know that the EPA sets the pesticide limit a hundred times lower than the smallest dose that causes any sign of harm to animals when fed that amount every day for most of their lives. In addition, rinsing apples well, making sure to rub the skin with hands or a fruit scrubber, helps remove pesticides.

In fact, apples—high in fiber and polyphenols, which function as antioxidants, and low in calories—are one of the healthiest foods. A diet rich in fruits and vegetables has been shown to help prevent stroke, cancer, and hypertension.

APPLES AND PESTICIDES

ANNUALLY the US Department of Agriculture (USDA) tests samples of fruits and vegetables for levels of pesticides. In the years when the USDA has included apples in its surveys, residues of pesticides have been found in a high percentage of the fruit, though most apples contained residues within the legal limits (pesticides are especially concentrated in apple juice imported from other countries). Children are at higher risk because their bodies are still developing and cannot handle toxins as well as adult bodies can.

Most experts maintain that fruits and vegetables are an important component of a balanced diet, despite the risk of ingesting toxins. Orchards that grow apples organically, with minimal chemicals, are more common today, and many people are willing to pay more for organic apples.

DID YOU KNOW?

- The **SCIENTIFIC NAME** for apple is *Malus domestica*.
- Apples are members of the **ROSE** family.
- **ARCHAEOLOGISTS** have found evidence that humans have been eating apples since at least **6500 BCE**.
- **PILGRIMS** planted the first apples in America in the Massachusetts Bay Colony.
- The average American eats **NINETEEN POUNDS** of apples a year.
- The **LARGEST APPLE** ever picked was three pounds, two ounces.
- To this day, most apples are harvested **BY HAND**.
- It takes the energy of **FIFTY LEAVES** to produce one apple.
- The science of apple growing is called **POMOLOGY**, from the Latin *pomum*, meaning "fruit."
- The number **FIVE** is important in the apple world. An apple blossom has **FIVE PETALS**, apple blossoms typically form in **CLUSTERS OF FIVE**, and each apple has **FIVE SEED CAVITIES**, creating the star you see when the apple is cut in half crosswise.
- The apple is the **STATE FRUIT** of Minnesota, as well as the state fruit of Illinois, New York, Rhode Island, Vermont, Washington, and West Virginia.

4

CORN

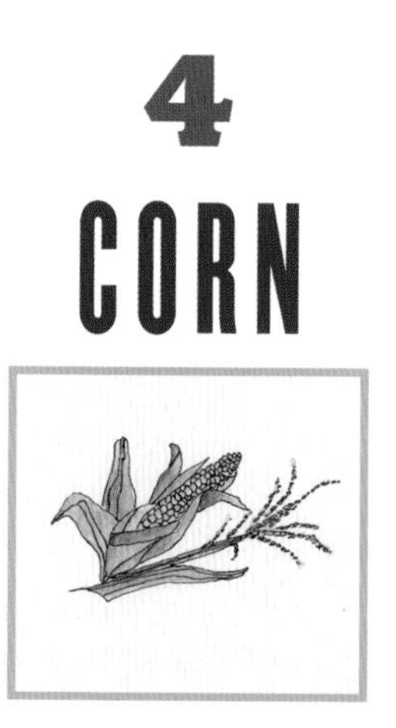

Heap high the farmer's wintry hoard!
Heap high the golden corn!
No richer gift has Autumn poured
From out her lavish horn!

JOHN GREENLEAF WHITTIER

Corn queens, corn festivals, corn feeds, corn mazes, corn dogs, and corn monuments (a corncob-topped water tower in Rochester, a twenty-four-foot corn sculpture in Olivia)—Minnesota pays homage to corn in many ways. And well it should. Corn is Minnesota's most profitable crop and covers the most acreage of all row crops. Indeed, it is the most widely grown grain crop throughout North and South America, with 332 million metric tons grown each year in the United States alone. It is very versatile, grows in many climates and soils, is easy to plant, and has numerous uses, culinary and otherwise. Thousands of corn varieties grow around the world, providing 21 percent of human nutrition and food for many animals. In the United States, the primary crop is dent, a variety with a high starch content. Minnesota is one of the top four corn producers in the nation, along with Iowa, Illinois, and Nebraska.

In 2014 Minnesota farmers produced 8.2 million acres of corn on approximately 25,000 farms across the state. That was a good year, with a harvest of a record 1.18 billion bushels of corn and an average of about 156 bushels per acre. The total crop was worth approximately $4.7 billion.

The path from corn's beginnings to the rich harvest of today has been a long one. The grain, which most of the world calls *maize* (*Zea mays*), is indigenous to the Western Hemisphere. Fossils of maize

Queen Gail Page, Cokato Corn Carnival, 1959. MNHS collections

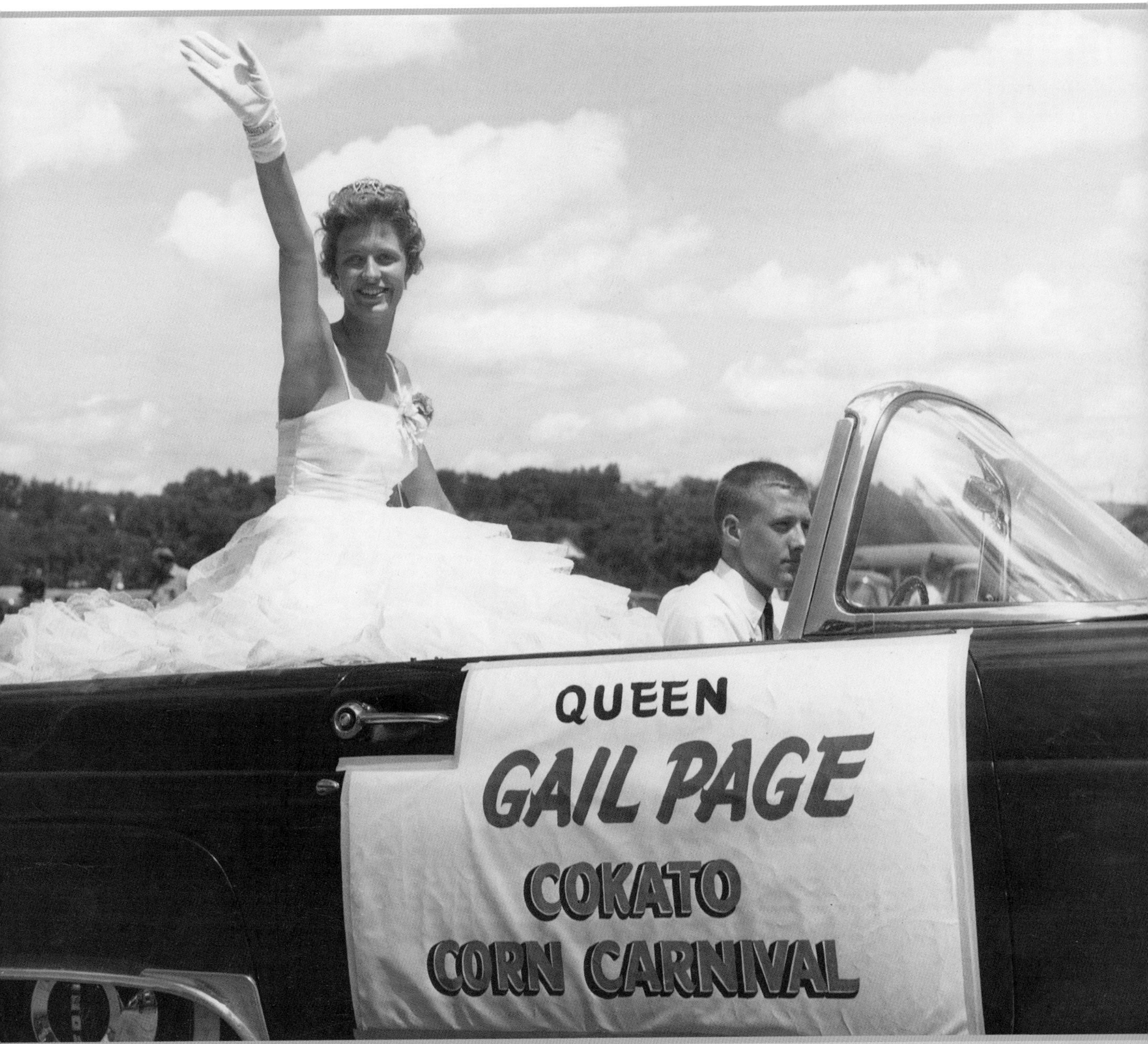

dating back 10,000 years have been found in the subsoil of Mexico City, and researchers there have found stone grinding tools covered with maize starch.

Over generations, humans have made small changes in corn's genetic structure, altering the plant's shape and taste. The domestication of maize was an impressive accomplishment for peoples living in small groups and often shifting settlements seasonally. Teosinte, a small, grassy wild plant that was maize's predecessor, has numerous thin branches, each bearing a seed enclosed in a hard shell. Early peoples were able to take this small grass with unwanted, inconvenient features and transform it into a high-yielding, easily har-

Minnesota cornfield. David L. Hansen, University of Minnesota

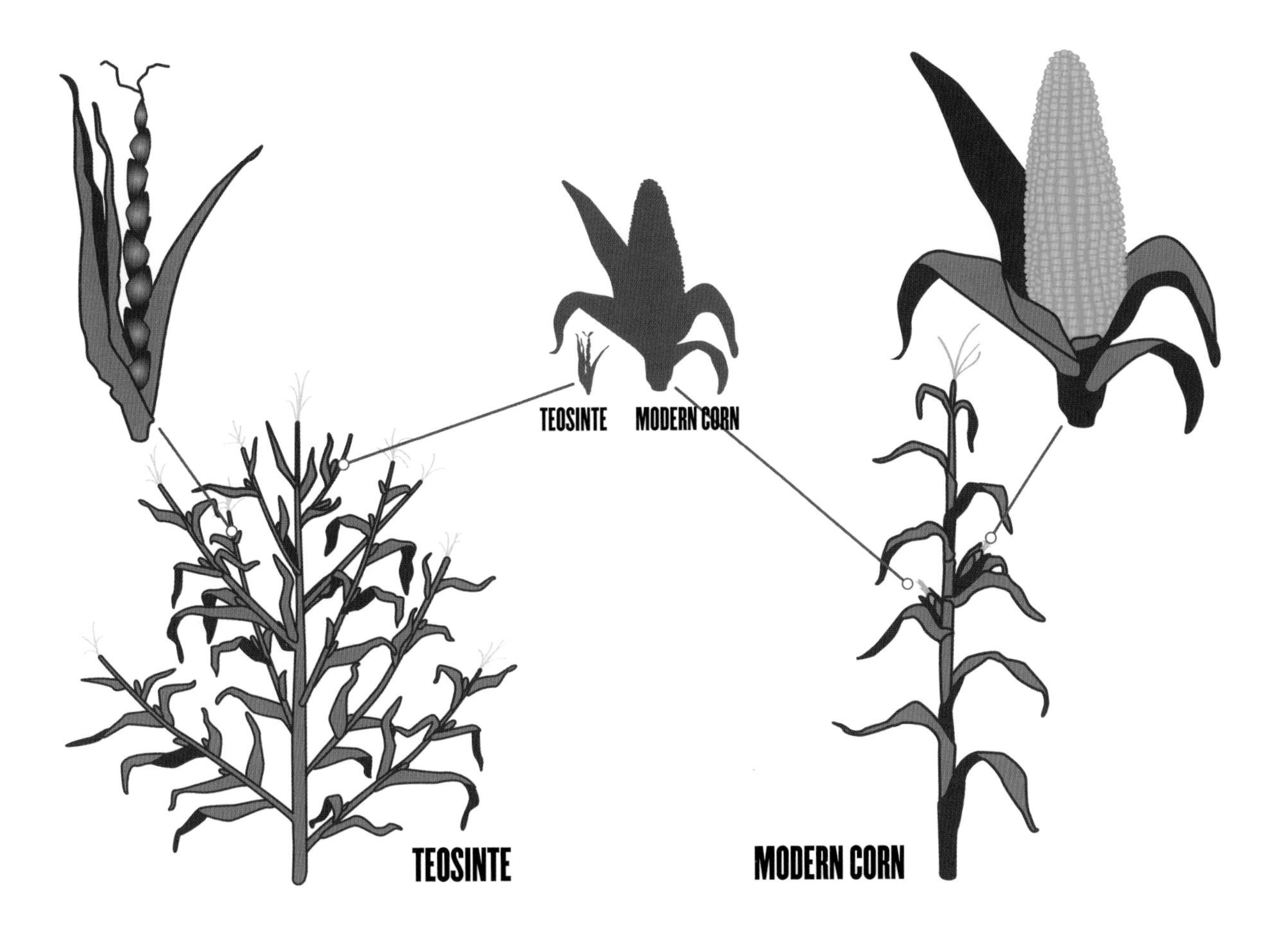

Teosinte was the precursor of corn. Graphic by Jennifer Osborne Anderson

vested food crop. The process of domestication probably occurred in stages over a period of several hundred years. Modern corn is perhaps the most thoroughly domesticated crop, and for centuries it has been completely dependent on humans for its existence. Were it not planted, tended, and stored by people, it would become extinct in several years' time.

The Maya, Incas, Olmec, and Aztecs cultivated numerous varieties of corn throughout Mesoamerica, choosing ones for cooking, for grinding, and for drying. They soaked corn in an alkali solution, usually lime water, to soften it and increase its nutritive value and flavor. This method, called nixtamalization, is still used today for corn tortillas, tamales, hominy, and many other items. The high yield and swift mutation of corn enabled these civilizations to develop rapidly, creating ingenious farming methods and complex cultures.

To the Mesoamericans, maize was not only a highly successful crop but it was also embedded in their culture through foundational

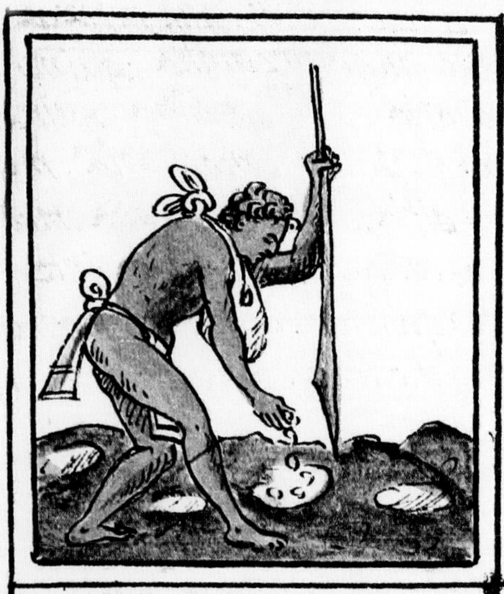

stories and religious rites. They celebrated its harvest with great festivities. Denoting maize's strength and importance, the Maya creation story, the Popol Vuh, states, "The first man was made of clay and was destroyed by a flood. The second man, made of wood, was swept away by a great rain. Only the third man survived. He was made of maize." Indeed, of all the grains eaten in the world, corn (or maize) probably is surrounded by more legends and folklore than any other.

Corn was such an important part of many cultures that its origin has often been framed in myths—sacred stories that embody a culture's values. For the Ojibwe, corn arrived as a gift. A young hunter named Wunzh took part in a ceremonial fast to learn about this spirit guide. During the three days of his fast, he wrestled with Mondawmin, who descended from the sky. On the third day, Mondawmin surrendered, asking that his clothing be taken from him so that he could be buried naked in the earth. He told Wunzh to keep the spot free of weeds and to place fresh soil on the spot monthly. The boy did as he was told, and in time he saw green plumes poking through the earth. Finally, Wunzh brought his father to the spot, and they both saw a graceful, tall plant with golden arms and bright, silken hair. Wunzh realized what had happened, saying, "It is my friend, Mondawmi. We need no longer rely on hunting alone. For this I fasted, and the Great Spirit heard."

Beginning around 2500 BCE, the crop spread through the Americas, probably by the diffusion of

Aztecs planting corn, from the Florentine Codex, a sixteenth-century Mesoamerica ethnographic research project by Franciscan friar Bernardino de Sahagún. Wikimedia Commons

seeds along trade routes. It moved first from Mexico to the southern United States about 4,100 years ago. Ten centuries before our time, the Anasazi, ancestors of the Hopi Indians of the Southwest, built their famous pueblo, a single enormous building that covered an entire settlement. The Anasazi grew a high-yielding variety of maize in their fields and made dough for tacos, which were cooked on flat stones and then rolled up before being eaten. The Adena culture, made up of peoples who lived from 800 BCE to about 1 CE and were displaced by the Aztecs, seem to have brought maize with them to their sites in the central Ohio Valley.

Maize was a part of the intensive agriculture of the Hopewell culture, which flourished along the rivers of the northeastern and midwestern United States from 200 BCE to 500 CE. It was also important to the Mississippian culture (800—1600 CE) in the

MILPA AGRICULTURE, ALSO KNOWN AS THE THREE SISTERS

MILPA IS A CROP-GROWING SYSTEM used first throughout Mesoamerica and then by native peoples across the Americas. The word "milpa" is derived from the Nahuatl (Aztec) word phrase *mil-pa*, meaning "to the field." Based on the ancient agricultural methods of Maya and other Mesoamerican people, the milpa method produces maize, beans, and squash. In a combination that came to be known as the "three sisters," corn is planted on a mound of soil, beans grow up the stalks, and squash (pumpkins) grow around the base. The beans replenish the soil with nitrogen, which the corn has removed. The squash hold down the weeds.

With the milpa process, fields are cultivated for two years and left fallow for eight. Agronomists point out that the system is designed to produce relatively large yields of food crops without the use of artificial pesticides or fertilizers. In the estimation of H. Garrison Wilkes, a maize researcher at the University of Massachusetts in Boston, the milpa "is one of the most successful human inventions ever created."

In Minnesota, public examples of the three sisters garden have been planted at the Science Museum of Minnesota, at Grand Portage National Monument on the North Shore of Lake Superior, and at the Fond du Lac Tribal and Community College in Cloquet.

Mississippi River Valley and points east and north. These people grew crops that intermingled corn, squash, and beans. So successful was their agriculture that it led to an explosion of population and the development of sophisticated cultures. The earliest archaeological evidence of human habitation in Winona County comes from a site, circa 800 BCE to 900 CE, known as the LaMoille Rock Shelter, south of the present city of Winona on the Mississippi. The people who lived there were agriculturists, growing crops of corn, beans, squash, and tobacco. Two other major sites of the Mississippian tradition (900–1200 CE) have been found near the junction of the Cannon River and the Mississippi River in the vicinity of Red Wing.

DNA studies show that throughout this period peoples continued to purposely select corns for certain traits. For example, in pre-Columbian Peru, thirty-five varieties have been found, including popcorns, flint varieties, corns for dying textiles, and some varieties for making flour. Corn today is typically yellow, but native peoples bred several colors (red, black, blue, white), as well as multicolored ears. As corn moved out of Central America, it became part of existing agricultural traditions. For example, in the eastern part of North America, it joined the complex of pumpkins, sunflowers, and lamb's-quarter. (Lamb's-quarter is today often considered a weedy plant, but it was cultivated in the past in Europe, India, and North America.) As it was introduced to new cultures, new varieties were developed and selected to better serve local diets and uses. In the northern states flint corn was valued for its ability to thrive in a short, cool season.

When Europeans first arrived in what is now Minnesota, they observed Dakota people growing corn and other cultivated crops. In the book *The Aborigines of Minnesota*, Professor N. H. Winchell, a geologist with the Minnesota Geology Station, wrote: "Throughout the area [Minnesota] of the United States, the aborigines, when first visited by Europeans were found to be cultivators of the soil. The earliest travelers found corn, beans, squash and tobacco in general use and derived from more or less regular crops . . . the industry of agriculture was . . . mainly in the hands of the women."

During the summer months Dakota families gathered in villages. The men hunted and fished, while women and children cultivated corn, squash, and beans. The digging tools were made of flat stones with handles attached. With only a slight disturbance of soil, the seeds could be planted in hills. The Indians saved the best corn seed each year for the next year's crop. Once the corn was harvested and stored, families focused on harvesting wild rice.

Dakota woman and children protecting corn, 1862. MNHS collections

Early observers noted that festivities accompanied the growing of corn. In August 1820, Michigan governor Lewis Cass and Henry Rowe Schoolcraft, an American geologist, geographer, and ethnologist studying Native American culture, were witnesses to a corn festival. Here at Little Crow's Dakota village, on the Mississippi near what is now St. Paul, Cass and Schoolcraft were invited to attend the annual "green corn dance." As described by Schoolcraft in *Narrative Journals of Travels through the Northwestern Regions of the United States*, the event proceeded thusly:

> While these things were going forward, the Indian women were busily engaged in gathering green corn, and each one came into the centre of the chief's cabin and threw a basket full upon a common pile, which made a formidable appearance before the speakers. . . .
>
> Our attention was now drawn by the sounds of Indian music which proceeded from another large cabin at no great distance. . . . Our curiosity, however, being excited, we applied to Governor Cass to intercede for us, and were by that means admitted. The first striking object presented was two large kettles full of green corn, cut from the cob and boiled. They hung over a moderate fire in the centre of the cabin, and the Indians, both men and women, were seated in a large circle around them. . . . The utmost solemnity was depicted upon every countenance not engaged in singing. . . . Suddenly the music struck up, and the singing commenced. In the course of these ceremonies a young man and his sister, joining hands, came forward towards the centre of the cabin. We were told they were about to be admitted to the rights of partaking of the feast. . . . When this ceremony ceased, one of the elder Indians, dished out all the boiled corn into separate dishes for as many heads of families as there were present, putting an equal number of ladles full into each dish. Then, while the music continued, they, one by one, took up their dishes and retiring from the cabin by a backward step, so that they still faced the kettles, separated to their respective lodge, and thus the ceremony ceased.

The Ojibwe, too, grew corn, squash, and pumpkins after they had moved to their summer homes. Women, who along with children and grandchildren were the primary farmers, planted corn in hills, with squash planted around the perimeter. After the corn was harvested, it would be roasted in the husks, parched in a skillet or kettle, or dried for later use.

THE YANKEES AND EUROPEANS ARRIVE

Many Euro-Americans arriving in Minnesota took up farming. Initially, agriculture grew slowly in the territory, first in the trading posts and then on individual farms. After Dakota and Ojibwe people were forced onto reservations, it developed more quickly as land was opened for purchase, as railroads were constructed, and as immigrants located good farmland. The Homestead Act of 1862, which made cheap land available, brought in many farmers eager to own land.

Corn was an early and essential crop. The initial farming area for Europeans probably was southern Washington County. The first farmers there were Joseph Haskell and James S. Norris, who set up their claim near Afton in 1840. That spring they turned three acres of sod in six days and planted the plots with corn and potatoes. Ten years later, the *Minnesota Pioneer* related that in 1850 Mr. Wm. Middleton in southeastern Minnesota was raising ninety bushels of "Indian" corn per acre. Recommended varieties for the state at the time were Golden Sioux, Northern Yellow Flint, Eight-Rowed Yellow, Canada Early White, Tuscarora, White Flour, and Rhode Island White Flint. Some farmers also planted yellow and white dent corn, high in soft starch. A few adventurous farmers tried corn breeding experiments during the 1850s to develop a crop better suited to Minnesota's climate and soil. In 1850 farmers harvested 16,725 bushels of Indian corn. By 1860 the harvest was almost 3 million bushels.

Corn was important because it could be planted in the newly broken, roughly prepared earth. Often the seeds were inserted into an opening made in the sod with an ax. Seeds could also be planted around tree stumps, after the trees had been felled but while the stumps remained. In early days farmers had few farm implements, usually a hoe, a spade, sickles, and a breaking plow. An account from the 1850s noted that corn was dropped by hand, covered with a hoe,

WHAT MINNESOTANS HAD TO SAY ABOUT CORN

Corn is the backbone of the farm economy and other businesses.

J. WILLIS W.

and plowed with a shovel plow. The weeds in the field were kept down by hand hoeing. When the corn was picked, the stalks were cut by hand and then gathered into shocks to allow them to dry and ripen. Most farmers ground their own corn, devising creative ways to accomplish the task. Some used a burned-out stump and a block of hardwood as a mortar and pestle. One placed his corn in a trough and rolled over it with a cannonball.

Between 1830 and 1860, many advances in farm machinery were made, often related to corn cultivation. In 1860 alone, the US Patent Office issued patents for corn shellers, corn huskers, corn cultivators, cornstalk shocking machines, seed drills, corn and cob mills, and hundreds of corn planters. Still, farmers were limited in what they could afford to purchase and how large their farm operation was. For many, the crop was labor intensive from planting to harvest, with everyone in the family needing to help out. And at the beginning of the Civil War, the patent office strongly advised farmers not to invest in "newfangled" corn planting machinery, since the job could be handled more cheaply and efficiently by the handwork of women and children.

WHAT MINNESOTANS HAD TO SAY ABOUT CORN

Corn is the major economic crop of Minnesota for grain, milling, oil, ethanol, plastics, high-fructose corn syrup, and many other useful products beside silage and grain for cattle.

DANIEL P.

In the nineteenth century and well into the twentieth, Minnesota farmers collected seed from their best plants for the next year's crop. Though the seed choices were somewhat tailored to local soils and climates, the plants matured unevenly. The stalks often fell over and broke off. From the 1860s until the 1920s, an open-pollinated variety was used, which yielded twenty-five to thirty bushels an acre. At the end of that era, the average price for corn was $.30 per bushel, and Minnesota had roughly 4.5 million acres planted, resulting in a value of $34.8 million for the state.

CORN RESEARCH AND MECHANIZATION

In 1893 Professor Willet Hays at the University of Minnesota selected and developed an early flowering strain that became known as Minnesota 13. It was better suited for cold climates and the short growing season and soon after its release in 1897 became the most

The corn shock was a storage method designed to shed rain reasonably well and prevent spoilage among the drying corn. Note the pumpkins amid the shocks of corn. The date on this photo from McLeod County is 1915, but it could have been fifty years earlier, so little mechanization had taken place in the intervening years. MNHS collections

popular corn variety in the northern states. Today about one-third of the seed planted in the United States can be traced back to Minnesota 13, with 45 percent coming from all the strains developed by University of Minnesota researchers.

The 1920s ushered in an era of research to develop hybrids that produced greater yields. This was accomplished by crossbreeding corn traits. The University of Minnesota researchers evaluated corn for numerous qualities, including resistance to insects and disease, date of maturity, early-season vigor, yield, and plant height. The St. Paul campus and the University Agricultural Experiment Station in Waseca became the centers for these experiments. As a result, the university released several hybrid corn varieties in the late 1920s,

CORN
THE FOOD OF THE NATION
SERVE SOME WAY
EVERY MEAL
APPETIZING
NOURISHING
ECONOMICAL
GRITS
CORN
MEAL
HOMINY
Lloyd
Harrison
UNITED STATES FOOD ADMINISTRATION

WHAT ABOUT GENETICALLY MODIFIED CROPS AND THE 10 PLANTS?

FIVE OF THE 10 PLANTS have been genetically modified (GM), with varying levels of acceptance by farmers and consumers. GM apples, alfalfa, corn, soybeans, and turfgrass have all been approved for sale in the United States. But each plant has been modified for different reasons and with different methods, and just as each plant has a unique growth cycle and characteristics, so each must be considered separately when attempting to understand GM plants.

More than 90 percent of the corn and soybeans grown in Minnesota are GM. Along with alfalfa, corn and soybeans are modified to allow herbicide applications of glyphosate (Roundup Ready®), resulting in more herbicide use, where the yield benefits can offset the higher herbicide and GM seed costs. GM alfalfa is a perennial plant and has the risk of persisting in the environment, unlike the annual crops of corn and soybeans. Most farmers feel these GM plants are superior and worth the additional cost.

Even with this high percentage of GM soybeans, Minnesota is still the top exporter of soybeans in the United States, and millions of non-GM soybeans are grown here for countries such as China. By 2016, both China and the European Union had approved importing GM soybeans and corn for food and animal feed, but not for planting/growing in their countries. Reports from the National Academies of Sciences, Engineering, and Medicine conclude these GM crops are safe for consumption, but controversy still exists. The *New York Times* reported that GM crop yields had not increased to levels anticipated and pesticide use had actually increased in some cases. Monsanto, the company that makes most of the GM seed, has disputed this assessment.

GM apples, even with the modification of an enzyme that prevents oxidation or browning of cut fruit, has met with public resistance and is scheduled to be on the US market in 2017. GM turfgrass has not been adopted to a great extent; there is no GM wild rice or wheat to date; and only traditional plant breeding has been conducted with white pine and American elm.

Discussions surrounding GM plants are complex and often challenging to understand, but it is important to analyze each crop separately because each plant is unique.

MINNESOTA 13

DURING PROHIBITION, many Stearns County farms accrued great profits from distilling corn liquor moonshine. Dubbed "Minnesota 13," the liquor became well known across America, said by many to taste very like Canadian Club. The moonshine was a premium-quality twice-distilled and properly aged whiskey. Minnesota 13 was the name of an open-pollinated corn, developed by the University of Minnesota and cultivated in Stearns County because of its shorter growing season. Many folks said that Holdingford was the "moonshine capital" of Minnesota.

Ultimately, federal agents were able to curtail these large operations by burning barns and sheds, by using informers to uncover facilities, and by increasing their surveillance. But for a decade, Minnesota 13 added to many farmers' incomes.

THE SOCIAL LIFE OF CORN

Corn is a friendly plant, one that people relate to, as the Ojibwe related to Wunzh. It stands as tall as a person and has a silken, hairlike top and leaves that wave like arms. Perhaps that's why folks have created so many stories and festivals that involve the plant. Minnesota is no exception. We have Corn Days, Corn Fests, Corn Carnivals, Corn-on-the-Cob Days, Corn Capital Days, and one Cornstalk Convergence. There are corn-husking contests, corn mazes, and corn-eating competitions.

Of course, much of this activity occurs because Minnesota raises so much corn. But farmers here grow soybeans, oats, and sugar beets too. The harvest of these crops, important as it is, does not create the same excitement. It would be hard to create a sugar beet maze or make bean "dolls." Corn's elegant, green stature is associated with beauty. Its growth and harvest are connected with abundance and the simplicity of life in an earlier time. By participating in one of the fall harvest events, people partake of a slice of farm life as well.

Many of these corn celebrations are not new. Morris had a Corn and Alfalfa Exposition in 1913. Ortonville hosted its seventy-seventh annual Cornfest in 2015. Cokato initiated Corn Days in 1950, and the Longfellow neighborhood in Minneapolis has had a corn feed for more than forty-five years. Some of these events are brief affairs; others, like Corn Capital Days in Olivia and the Cokato Corn Carnival, extend for days, with parades, arts and crafts, displays of vintage farm equipment, entertainment, food and drink, and fireworks. Cokato serves more than twelve tons of free corn on the cob to the thousands who attend its carnival. For many towns, the event functions as an annual reunion for current and former residents—a way to mark the passing of time and a way to remember traditions.

Husking bees may be a thing of the past, but there are still husking contests—regional, statewide, and national. Here contestants in various age and gender categories walk along rows of standing corn, pulling off ears as quickly as possible. They use a husking hook on

WHAT MINNESOTANS HAD TO SAY ABOUT CORN

Skeptics were proven wrong, and by the 1930s Minnesota, which had been deemed "unsuitable for growing corn," became part of the Corn Belt, and corn became Minnesota's most widely grown crop.

KURTIS G.

In 1907, Anoka created a Corn and Potato Palace between Second and Third Avenues as part of its corn festival. MNHS collections

one hand, tossing corn into a moving wagon. The husking is timed, and huskers are docked for any ears left on the stalk.

The first national husking contest was held in Alleman, Iowa, in 1924. Because this was a new event, the turnout that year was not large. However, each year the contest attracted more people and from a wider area. When Minnesota hosted the national event in November 1934, in Fairmont, Martin County, the contest attracted an estimated 75,000 people. Fairmont needed 800 policemen, 250 guardsman, and 400 legionnaires directing traffic. From nine o'clock to noon, so many cars were approaching from the west on Highway 16 that patrolmen on the scene made it a one-way road. Four planes hovered over the area, eventually landing near the contest. Small wonder the crowd was said to be almost "beyond description." Minnesota's Ted Balko of Redwood County was the winner with twenty-six bushels.

National corn-husking contestants had been entertained the previous night at a banquet at the American Legion Hall, "with dignitaries in attendance." The following day a parade, including eleven companies of the National Guard from neighboring towns and the American Legion Drum and Bugle Corps from Mankato, marched through the center of town. The contest was covered by several farm papers and broadcast by NBC radio, with a portable transmitter to convey real sounds to the listeners. It is no wonder that two years later, in 1936, *Time* magazine called corn husking "the fastest growing sporting spectacle in the world." Winners were national heroes, and some even received marriage proposals. The record was set in 1940 at forty-one pounds picked per minute.

Corn mazes dot the autumn landscape. This farm in Shakopee operates the oldest corn maze in the Midwest. Since the mid-1990s mazes have become popular family attractions and a source of income for farm families. Courtesy Sever's Fall Festival

One might wonder why the contests were so popular in the 1930s. Doubtless, part of the reason was that the country was in the throes of the Depression, and almost everyone picked corn by hand at that time. Few farmers had mechanical corn pickers. They understood that the work was hard and often slow and could appreciate a swift, skilled picker. Farm families needed diversions, and corn-husking competitions were free. In addition, along with the contests these events offered bands, nickel hamburgers, and a chance to share stories with other spectators.

Minnesota and the nation have continued this tradition through the years, with a hiatus during World War II and for the following two decades. When men came home from the war, they found that corn husking had become mechanized, and their skills were no longer needed. The competitions were revived in 1975 and have continued to the present. In 2015 the Minnesota competition was held at the Dwain Gerken Farm near Oak Center in Goodhue County.

CORN *and the* ENVIRONMENT

CORN, with its associated production and uses, is a complex subject that affects nearly all of us either directly as consumers or indirectly through its impact on the economy and the environment. Corn production has affected Minnesota's environmental landscape. Although Minnesota corn is produced primarily as a feed source for livestock, the public uses corn by-products daily. The most recognizable nonfood by-product might be ethanol, a legislated part of our state's fuel system, whereas the most recognizable food by-product might be high-fructose corn syrup, used as a sweetener in most soft drinks and beverages.

The economic impact of corn is formidable, with approximately 7 million acres grown annually in Minnesota. While the income from corn harvests is a major economic driver for the state, government policy and programs, including commodity subsidies and insurance, offer noteworthy incentives, often resulting in farmers growing more corn. The environmental impacts from corn production result from the inherent inefficiencies of nutrient management and nutrient fate. Minnesota's climate, soils, and complex surface and subsurface water interactions are key contributors to this environmental liability—along with farmers' risk management attitudes and practices. As a result, the number of corn acres in Minnesota continues to increase, with direct environmental consequences from nitrates in surface and subsurface water, which eventually contribute to hypoxia, or oxygen deficiency, in the Gulf of Mexico. In addition, phosphorus levels are also increasing in Minnesota's surface water as soil erosion carries phosphorus off the fields and into the water system.

Economic incentives in recent years have encouraged farmers to plant more corn, at the expense of perennial grasslands or other crops such as alfalfa. Because corn is an annual tilled monoculture, with nearly bare soil for more than half the year, there is concern about the environmental effects of such huge cultivated areas.

In recent decades, corn has expanded significantly from southern Minnesota northward into the Red River Valley. In Minnesota, corn receives the highest level of nitrogen fertilizer of any crop. Although soybean and alfalfa use more nitrogen than corn, nitrogen fertilizer is not commonly applied to these crops because as legumes they produce their own nitrogen from the atmosphere. Adding nitrogen fertilizer to corn acres to maximize yields results in greater nitrogen losses through movement with water or attached to soil. In Minnesota, the majority of corn is grown in the Minnesota River basin, which is part of the Mississippi River watershed that empties into the Gulf of Mexico.

Nitrogen mostly moves throughout Minnesota surface and subsurface waters as nitrate, although the majority of nitrogen fertilizers are ammonium based. In soil, over time, ammonium-based nitrogen converts to nitrates, which affect Minnesota's lakes, smaller rivers, and streams. A ten-year study by the Minnesota Pollution Control Agency (MPCA) found that 70 percent of the nitrates in these waters come from production, or cultivated row crop, agriculture. Fertilizer applied to cornfields is not the only source of these nitrates; other sources include direct runoff of manures and the discharge from some sewage treatment plants. In the southern portion of Minnesota, an MPCA report found that 27 percent of the state's lakes and rivers contain

continued next page

nitrate concentrations that exceed the drinking water threshold for nitrates.

Economic incentives to grow (more) corn in recent years are also contributing to farmers forgoing practices that traditionally favored the environment. When legumes (e.g., soybean, alfalfa, clovers) were part of traditional crop rotation, as they produced their own nitrogen from the atmosphere there would be "leftover" nitrogen to benefit the following crop. Rotations also provided a diverse pest community and plant rooting structures that influence soil water profiles, yet the economics and related incentives of growing continuous corn have resulted in fewer crop rotations today. Additional wildlife habitat has been lost from pastures, and tiled wetlands have changed the movement and quality of water.

Corn producers, and the entire corn industry, have made numerous positive strides in protecting the environment. Corn breeding efforts have improved the efficiency of nitrogen use and introduced hybrids that are better adapted to local environments. The fertilizer industry has developed products that attempt to keep nitrogen fertilizer in the ammonium form rather than in the nitrate form. Corn farmers have also changed some of their management practices to ones that benefit the environment through conservation tillage to reduce soil erosion and phosphorus movement to surface waters. Farmers are also gradually delaying fertilizer applications to better coincide with nutrient usage by the plant.

Currently, four watersheds in the state are taking part in the $9.5 million Minnesota Agricultural Water Quality Certification Program, begun in 2013, to help farmers who are willing to adopt precise planting methods for protecting water from agricultural pollutants. Additionally, nine agencies are working with Minnesotans to create a statewide strategy to reduce nutrients in the state's waters. The 2014 nutrient reduction strategy outlined by these groups suggested the following changes for Minnesota farmers: increase fertilizer use efficiencies; increase living cover crops; minimize field erosion by leaving crop residue or provide grassed waterways and structural practices for runoff control; and improve tile drainage water quality treatment and storage.

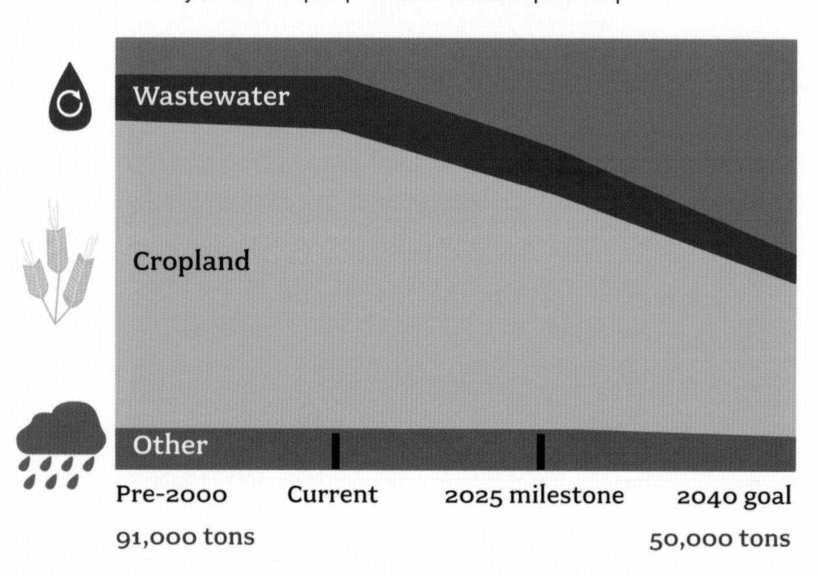

Corn production is a complex issue that has a huge impact on farmers, consumers, and the environment. Recognizing the trade-off of growing a high-income crop that requires careful management and perhaps regulatory enforcement of the use of nitrates may be necessary to ensure clean water for the future of Minnesota and the waterways that flow from our state. Graphic by Jennifer Osborne Anderson

DID YOU KNOW?

- The **SCIENTIFIC NAME** for corn is *Zea mays*.
- Corn's **WILD ANCESTOR** was *Zea mexicana*, which originated in southeastern Mexico and Guatemala.
- Corn was domesticated **10,000 YEARS AGO**.
- Christopher **COLUMBUS MENTIONS CORN** in his journal, saying it was "well tasted." Later he took corn with him when he returned to Europe.
- The original **"CORN PALACE"** was built in 1892 in Mitchell, South Dakota.
- The average ear of corn has **800 KERNELS**, arranged in sixteen rows.
- There are more than **4,000 USES FOR CORN**.
- **SEVENTY-FIVE PERCENT** of all grocery items contain corn in some form.
- Corn is grown on every continent of the world **EXCEPT ANTARCTICA**.
- **OLIVIA**, in Renville County, is known as the **"CORN CAPITAL OF THE WORLD,"** and Renville County produces the most corn of any Minnesota county.
- The value of the US corn crop in 2012 was more than **$85 BILLION**.

5
LAWNS AND TURFGRASS

The lawn should be a broad, uninterrupted expanse of grass, for it is the canvas on which we create our landscape picture.
C. GUSTAV HARD, UNIVERSITY OF MINNESOTA EXTENSION AGENT, 1958

Imagine a house without a lawn, a golf course without smooth greens, a baseball game without a groomed field. Indeed, they're all hard to picture. Lawns and turf are so much a part of our lives that we cannot imagine a different scene. Yet the lawn as we now know it—emerald green grass, uniform blades, weed-free—is a relatively recent obsession. Perfectly manicured golf courses and carefully tended sports fields, now standard practices, were scarcely dreamed of in the early twentieth century.

Earlier in America the area outside a home's front door might have had packed dirt or a mixture of flowers, herbs, and vegetables, or simply weeds. Our native grasses were unsuitable for a tidy lawn, and our sporadic rainfall was insufficient to keep grass green all summer. Two of our first presidents, Washington and Jefferson, resorted to using sheep to keep their lawns short. Even President Woodrow Wilson, in a show of support for the troops during World War I, had sheep brought to the White House front lawn to mow the grass. By such means he saved manpower and earned $52,823 for the Red Cross by selling the animals' wool.

England, with its gardens and lawns, became the model. During the 1800s, well-to-do Americans traveled abroad and observed the lush, manicured English lawns, which were easily obtainable with the cool, temperate climate and consistently moist conditions. Travelers brought back dreams of installing such lawns here. Unfortunately, America's harsh climate and uncertain rainfall made these lawns difficult to replicate, especially without a garden staff to scythe, weed, and water the grasses. Only the wealthy could afford such luxuries.

The ideal lawn has smooth green grass and is weed-free. University of Minnesota Extension

In "April," from Hours of King Henry VIII, *illustrated by Jean Poyer, the grass is filled with small flowers.* The Pierpont Morgan Library, New York

Although lawns are relatively new in America, the concept is an ancient one. The human preference for a lawn—or clear space—in front of our homes may be innate. Studies have shown that we prefer an environment that provides "prospect and refuge." Humans want to feel safe and to be able to observe the area around them. In the past people might have asked, "Are wild animals approaching?" or "Are the people coming toward us friendly or hostile?" Scientists point to Africa many thousands of years ago, when our ancestors surrounded themselves with low-growing savanna or grasslands.

Such clear, open spaces have been incorporated into landscapes in many forms. In medieval Europe, grasslands were maintained around castles so that guards had an unobstructed view of approaching visitors. Village commons, or *laundes* (a Middle English word), referred to the meadows held "in common" for villagers to graze their sheep and cattle; the hooved animals kept the area mowed. Early medieval lawns, called "flowery meades," were meadows studded with flowers and kept short by use. They might be filled with columbine, borage, campion, primroses, violets, wild strawberries, and daisies as well as common grass. In the sixteenth century the wealthy of Europe cultivated lawns for beauty. Such areas were generally planted with chamomile or thyme rather than grass. Closely shorn grass first emerged in seventeenth-century England at the homes of prosperous landowners, who often used sheep to keep the grass shorn.

Many Victorians valued a smooth swath of green to set off the house and garden and serve as a background for croquet or tennis. Well-known American landscape architects like Andrew Jackson Downing and Frederick Law Olmsted lauded the aesthetic qualities of a broad expanse of turf. In one of the first books on lawn care,

TABLE 5—1: SIZE OF THE LAWN CARE INDUSTRY (ESTIMATES)

ACRES OF HOME LAWN IN US
25 million (a little less than half the size of Minnesota)
ACRES OF TURFGRASS IN US
50 million
NUMBER OF EMPLOYEES IN THE INDUSTRY
1.6 million
MONETARY VALUE IN US
$60 to $75 billion
NUMBER OF LAWNMOWERS IN US
38 million
NUMBER OF GOLF COURSES IN US
15,753

Source: University of Florida IFAS Extension.

Fertilizing a campus lawn.
iStock.com/groveb

master research partnership with the US Golf Association. The purpose of this initiative, called Science of the Green, is to study and develop solutions to golf's present and future challenges.

Much of this research on sports and commercial areas has been applied to home lawn care. Equipment improvements (sprinklers, spreaders, seeders), grass hybridization, and chemical innovation have also stimulated the lawn industry.

Today lawns and turfgrass are a multibillion-dollar business, with about $60 to $75 billion spent in the United States each year. Turfgrass is the third-largest crop in total acreage nationwide; in Minnesota alone, approximately 700,000 acres of land are devoted to lawns. The turfgrass industry (sod farmers, lawn care workers, seed companies, retail and wholesalers, golf course greens managers, and so forth) contributes $8 billion yearly to the state's economy.

Braemar Golf Course, with its perfect fairways. Photo by Peter Wong, courtesy City of Edina

Certainly turfgrass has many benefits. It provides a safe and resilient foundation for some of our favorite outdoor activities. It adds beauty to our homes, setting off garden and house in a green frame. Grassy areas can decrease noise by eight to ten decibels, especially in an urban area. A healthy lawn protects the soil from water and wind erosion. As a perennial plant, turfgrass adds organic matter to the soil as the roots and other plant parts annually die out and decompose. Turf absorbs heat, reducing the need for air conditioners, and also adds oxygen back to the air through photosynthesis.

As more and more of our population resides in urban and suburban areas, more green space is covered with paved and other impervious surfaces. Runoff from these hard surfaces such as parking lots and roads contains many contaminants. Turfgrass purifies the water as it leaches through the root zone and into underground aquifers. The millions of grass plants in lawns, golf courses, and the like help clean the air, trap dust, and remove carbon dioxide and other greenhouse gases from the atmosphere.

TIPS FOR A HEALTHY LAWN

MANAGE IRRIGATION PRACTICES TO PREVENT OVERWATERING. Mandatory rain or moisture sensors that automatically shut off a sprinkler system when it rains are good solutions to preserve water. Do not allow water to run off into streets.

START WITH A SOIL TEST. Get your soil tested before applying fertilizer to see what your soil actually needs. This step could save both time and money.

RETURN LAWN CLIPPINGS TO YOUR YARD. Lawn clippings benefit the lawn, providing nutrients and shading the surface, thereby reducing moisture loss.

CONVERT SOME TURF TO "NO MOW" LAWNS OR OTHER PLANTS. Evaluate how much lawn you really need and consider transitioning some areas to native grasses, wildflowers, shrubs, or perennial flowers.

KNOW YOUR MOWING HEIGHT. In the summer you can improve your lawn's stress tolerance by increasing the mowing height by half an inch. For most lawn areas, mowing at a height of 2.5 to 3.0 inches will provide a good-quality turf.

USE A PUSH MOWER. If you have a city lot or a small yard, try a push mower. The new ones are easy to use and quiet and come with many options.

For more information on proper lawn maintenance, visit http://www1.extension.umn.edu/garden/yard-garden/lawns/.

As mowing height decreases, the depth of grass roots also decreases; home lawns are more resilient to drought and pests when cut at 3 to 4 inches. Graphic by Jennifer Osborne Anderson

TURF *and the* ENVIRONMENT

A NUMBER OF ENVIRONMENTALISTS, including experts and passionate citizens, have expressed concern over Americans' obsession with lawns. Weed-free and always perfect lawns can require 54 percent more water than mixed landscaping of trees, shrubs, and perennials. Mowers and other lawn equipment contribute to noise pollution, and a typical (four-horsepower) mower operating for one hour produces as much smog-forming hydrocarbons as an average car does in traveling a hundred miles. In addition, the runoff of nutrients from fertilizers and herbicides is considerable. For example, phosphorus became such a serious problem that in 2002 the state legislature passed the Minnesota Phosphorus Lawn Fertilizer Law, restricting phosphorous in fertilizer to be used on lawns. The law has benefited lakes and rivers.

Many of these problems could be mitigated with appropriate environmental practices. In 2011 the Environmental Protection Agency (EPA) passed new rules requiring gas-powered engines in lawn and garden equipment to cut smog-forming emissions by 35 percent. The EPA estimates that once these rules are fully implemented, annual emission reductions of hydrocarbons, nitrogen oxide, and carbon monoxide will be significant.

Homeowners themselves could obtain a soil test before applying fertilizers to see what their lawns actually need. Many chemicals are applied routinely by the season whether or not they are necessary. As an alternative to chemicals, lawn clippings provide several benefits when left on the lawn. They shade the soil surface and reduce moisture loss due to evaporation. In addition, decomposing grass blades are a valuable source of plant nutrients, especially nitrogen, and can reduce the need for supplemental fertilizer.

To help conserve water, homeowners can water infrequently but thoroughly during times of drought. They should avoid frequent watering, which promotes shallow grass roots and encourages weeds. Watering early in the day minimizes evaporation. Water only when it hasn't rained for at least seven days and soil moisture is low.

The Sustainable Urban Landscape Information Service (SULIS), part of University of Minnesota Extension, has as its mission "to enable homeowners, business owners, and horticultural professionals to create outdoor spaces that are functional, maintainable, environmentally sound, cost effective, and esthetically pleasing." To that end, SULIS has developed a series of online pamphlets on sustainable lawn care, at http://www.extension.umn.edu/garden/landscaping/maint/benefits.htm.

Recently many homeowners have shown an interest in low-maintenance turfgrass, or "no mow" lawns. Breeding programs at universities are now looking for new types of grass; some are native to the United States and will remain green with minimal inputs of water or fertilizer. Most of these grasses are fine fescues that grow well in drier soils and need minimal fertilizer. Backyards, or areas that are less visible, are good locations for fine fescues that may be mowed less frequently. Fine fescue planted on the St. Paul campus of the University of Minnesota is mowed only twice a year.

ABOVE: Home lawn irrigation is a common sight in Minnesota; however, many lawns will survive well without in-ground irrigation systems. iStock.com/bozhdb.

LEFT: Low-maintenance fine fescues establish easily and grow slowly, requiring less mowing than other cool-season lawn grasses. Mary H. Meyer

DID YOU KNOW?

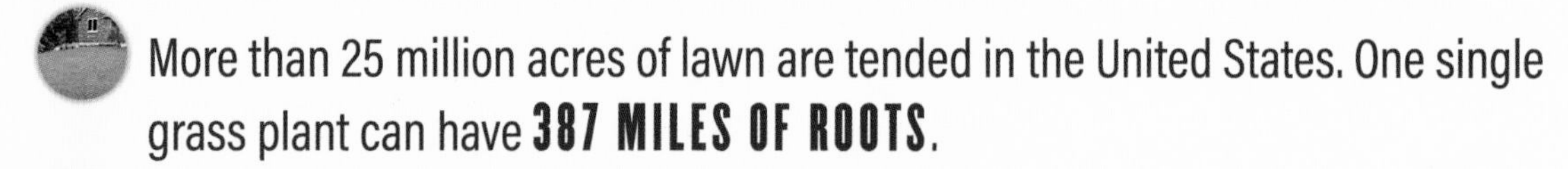

- More than 25 million acres of lawn are tended in the United States. One single grass plant can have **387 MILES OF ROOTS**.
- In a thick, healthy lawn there are 6 turfgrass plants in each square inch, 850 turf plants in a square foot, and about **7 MILLION TURF PLANTS** in an average lawn of 8,000 square feet.
- US turf traps an estimated **12 MILLION TONS** of dust and dirt annually.
- A fifty-by-fifty-foot lawn **RELEASES ENOUGH OXYGEN** throughout the day to supply the needs of one person.
- Grass typically grows **2.5 INCHES** in one week.
- Of the estimated **7,500 GRASS SPECIES**, 50 are cultivated for turf.
- For the **WORLD CUP SOCCER** matches in 1994, the Pontiac Silverdome grew its turf off-site, transported it to the dome in containers, and then reassembled it for the elite soccer matches. The Beijing Olympics in 2008 did the same.
- The very **LOW GROWING POINT** of grasses enables them to be cut and walked on without damage.
- Some gardeners enjoy cutting the lawn, likening the chore to **WALKING A LABYRINTH**, a type of meditation.
- The **FRAGRANCE OF CUT GRASS** makes us feel happy and relaxed.

6
PURPLE LOOSESTRIFE

A welcome European immigrant which usually grows in dense colonies, it rewards us for our hospitality with showy masses of color.

MINNEAPOLIS SUNDAY TRIBUNE, MAY 21, 1950

Beautiful, adaptable, hardy, fertile—all are words that describe the noxious invasive plant purple loosestrife (*Lythrum salicaria*). Indeed, these attributes account for loosestrife's rapid spread through Minnesota's wetlands. It is found in seventy-seven of the state's eighty-seven counties and much of Canada, all the Canadian border provinces, primarily in streams, rivers, ponds, and wetlands. According to the US Department of Agriculture's PLANTS database, purple loosestrife grows in every state in the continental United States except Arizona, Louisiana, Florida, Georgia, and South Carolina. At the peak of the loosestrife infestation, an estimated 300,000 acres of North American wetlands were affected.

Once established, loosestrife stands are difficult and costly to eradicate. Minnesota has been particularly vulnerable because of its large number of wetlands. But thanks to biological controls with the beetles *Galerucella pusilla* and *G. calmariensis*, Minnesota's waterways are being cleared of the nuisance plants.

The showy flower of Lythrum salicaria. iStock.com/Erik Agar

in what became the Eloise Butler Wildflower Garden and Bird Sanctuary in Minneapolis and recorded its presence in her log on July 15, 1916. Records show that in the 1920s several garden clubs in Minnesota planted purple loosestrife in the wetlands for "beautification." Its beauty was also mentioned in the May 21, 1950, edition of the *Minneapolis Sunday Tribune*. Along with a photo of Birch Pond from the wildflower garden, the paper noted, "The Purple Loosestrife grows as high as six feet in wet meadows and margins of pools. A welcome European immigrant which usually grows in dense colonies, it rewards us for our hospitality with showy masses of color like this."

In the April 1958 issue of the *Fringed Gentian*, the wildflower garden's publication, Martha Crone, the garden's curator, suggested planting loosestrife but hinted at its invasiveness: "It is a good plant to grow along streams, margins of ponds or in wet meadows. Especially where the competition is too severe for less aggressive plants to grow. The Plant is a long-lived perennial and produces graceful spikes of purple or pink flowers. They bloom during July and August. When once established it is hard to eradicate and will crowd out other weaker growing plants. It can be grown in garden borders where it remains smaller and does not spread."

During the twentieth century, nurseries developed a number of garden varieties of loosestrife, such as Morden Pink and Dropmore Purple, which were considered sterile and thus safe, noninvasive horticultural cultivars. Unfortunately, this assumption did not turn out to be correct.

THE PLANT TAKES OFF

When a plant from one continent is introduced to another, it often leaves behind the natural enemies that help control population explosion. Purple loosestrife was held in check in Europe and Asia by numerous insects that fed on it. In North America it had no natural pest and pathogen enemies.

The plant moved west from the Atlantic coast across the United States and Canada, most likely as modern highways were built. The soil disturbance of construction and maintenance helped loosestrife take hold in nearby drainage ditches and canals. Because of its appeal as a perennial garden plant, many individuals and nurseries also helped it spread.

The plant appears to thrive wherever excess nutrients are available, such as with the use of fertilizers for farming. The human activity of disturbing the land for crops and housing developments promotes the rapid growth of loosestrife. Once established, the plant can multiply quickly. The tiny seeds are easily spread by water, wind, wildlife, and humans. The plant has an extended blooming season, allowing it to produce enormous quantities of seed. Estimates of seed production rates range from just over 100,000 seeds per plant for young specimens with single stems to more than 2.5 million seeds per plant for established specimens with an average of thirty stems per plant.

Loosestrife is efficient at spreading vegetatively, forming dense, impenetrable stands. In addition, loosestrife seeds remain viable for up to ten years and can germinate with just ten minutes of sunlight. Loosestrife can produce new plants from any part that is cut and drops into water and can grow new leader shoots from whatever stalk is remaining. Loosestrife is tolerant of many soil types (gravel, sand, clay, or organic soil) and a wide range of environmental conditions (varied sunlight, pH, and temperatures). Taken together, all these factors enable the plants to colonize an area speedily.

The flower attracts long-tongued bees and butterflies, insects, honeybees and bumblebees, and cabbage white butterflies. Yet as the plant replaces native vegetation, wetland animals, including ducks, geese, toads, frogs, and turtles, lose their nesting sites, food sources, and cover. The plant also offers cover to foxes and raccoons, which prey on eggs and nests of other animals. The roots and overgrowth form dense mats that choke out native plant life. Loosestrife crowds out at least forty-four kinds of native grasses, sedges, and flowering plants that offer higher-quality nutrition for wildlife. Many songbirds and fish depend on these native plants that are displaced. Wetland bird species impacted by the infestations are the marsh wren, black tern, pied-billed grebe, and least bittern; unfortunately, these birds cannot eat the hard seed of loosestrife.

As the plant reduces habitat, it has a negative impact on fish spawning, especially in northern pike sites. Populations of muskrats and bog turtles have also been reduced, and areas where wild rice grows and is harvested are degraded by the presence of loosestrife.

Wetlands are the most biologically diverse, productive component of Minnesota's ecosystem. Hundreds of species of plants, birds, mammals, reptiles, insects, fish, and amphibians rely on healthy wetland habitat for their survival. The effect of purple loosestrife on native plant life in North America has been dramatic, with more than 50 percent of the biomass in some wetland communities having been displaced.

WHAT MINNESOTANS HAD TO SAY ABOUT PURPLE LOOSESTRIFE

It was planted as a landscape plant but proved ultimately to be an invasive species. This plant became a wake-up call to us, that we should be more careful of some plants from outside our area. We now have to fight to keep it (and other invasive nonnative plants) in control. Live and learn—that's the only benefit from this plant!

KATHERINE M.

In addition, loosestrife can invade drier sites. Concerns are increasing as the plant becomes more common on agricultural land, encroaching on farmers' crops and pastures. In urban areas, loosestrife can take hold in ditches and can block or disrupt water flow.

In January 2000, the US Fish and Wildlife Service declared purple loosestrife "public enemy number one" on federal land. In 1996 the Nature Conservancy ranked it number two in a survey of most troublesome weeds in wildlands.

ONE OF THE GREAT SUCCESS STORIES

Early control methods included physically removing plants, mowing, regulating water levels in marshes, and applying herbicides. These methods can be effective when the infestations are very small, but in larger populations the results from these efforts were disappointing. By the 1980s the Department of Natural Resources (DNR), along with concerned Minnesotans, took note of the proliferation of loosestrife stands along the state's many waterways. Several environmental and conservation groups organized the Purple Loosestrife Coalition in 1984. They approached the state legislature seeking a law prohibiting the sale, transport, possession, or planting of purple loosestrife. That bill was tabled, but it led to a new collaboration with a group of state and federal agencies: the DNR, the Minnesota Department of Transportation, the Minnesota Department of Agriculture, and the US Fish and Wildlife Service. This cooperative group conducted a study to identify the state's risks if the problem was ignored and then created a plan of action. In 1987 the state legislature passed a law making it a misdemeanor to sell purple loosestrife (*L. salicaria*). The Minnesota commissioner of agriculture designated the plant a noxious weed (Minnesota Statutes sections 18.75–18.91), which required landowners to remove or control the plants growing on their land.

In 1987 the DNR piloted the Purple Loosestrife Management Program with a $196,000 biennial budget funded by the Legislative Commission on Minnesota Resources. The management program

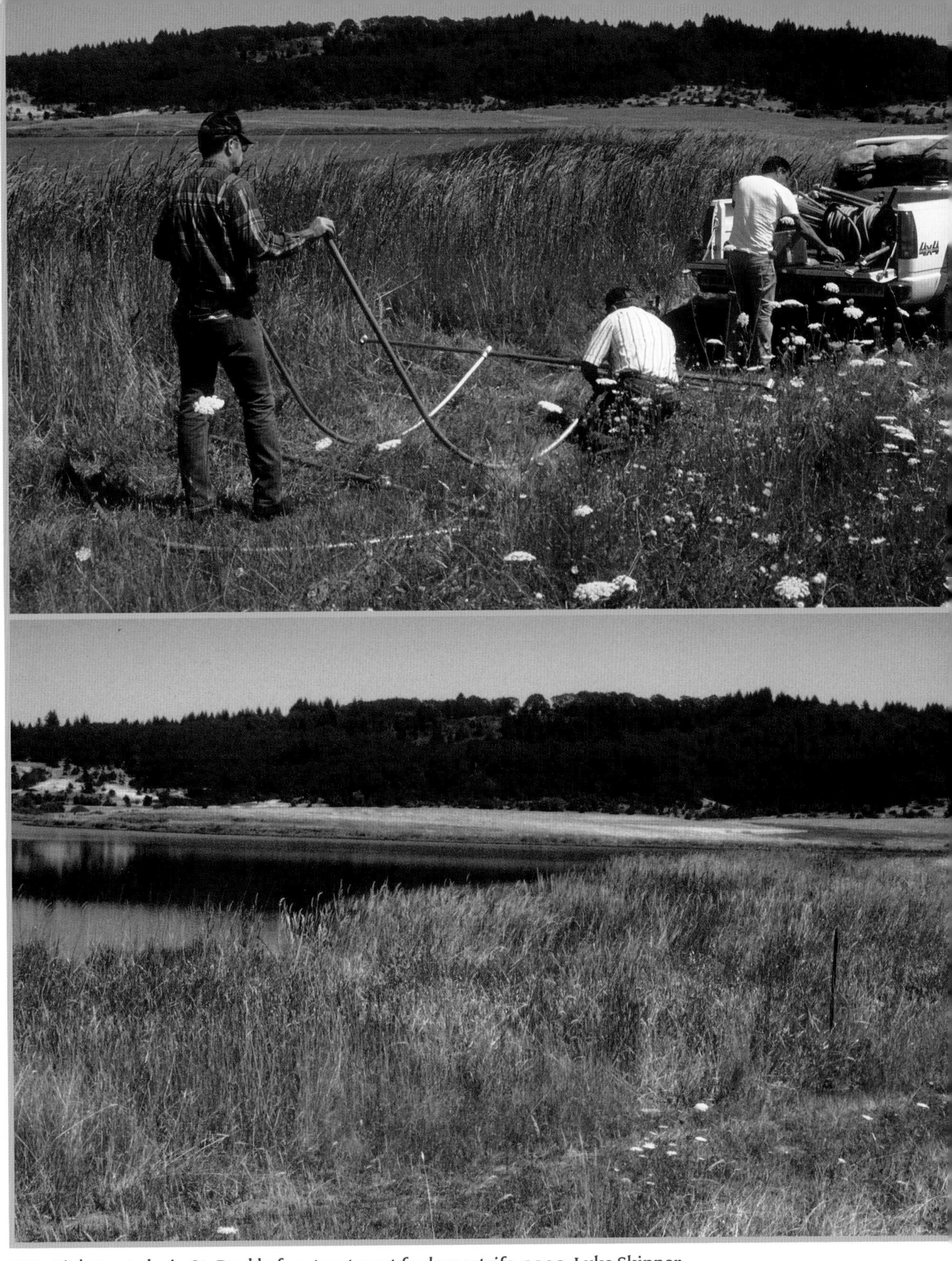

TOP: Pig's Eye Lake in St. Paul before treatment for loosestrife, 2000. Luke Skinner
BOTTOM: Pig's Eye Lake in St. Paul after treatment with beetles, 2004. Luke Skinner

included a broad public awareness campaign, an inventory of purple loosestrife infestations, research into control methods, and recommendations for action. The first efforts involved the use of herbicides to curb the stands. The DNR soon learned that herbicides do not result in long-lasting reduction of loosestrife when applied to a large, well-established population. Over time, as biological controls became available, the DNR gradually reduced herbicide use. Now staff use the chemicals only on small and new infestations, which have a limited seed bank.

Many individuals and nurseries thought the loosestrife cultivars they held were sterile ones that would not present problems for wild areas. However, research at the University of Minnesota showed that all the plants could cross with other loosestrife strains and grow quickly in the wild. As part of the education campaign, the DNR worked to persuade gardeners to remove the plant from home landscapes.

In the mid-1980s, when scientists began looking for biological controls, they turned to Europe, where the plant was kept in check by natural predators and pathogens. These biological studies covered 140 sites from central Finland to the Mediterranean coast. The researchers understood that utmost caution was necessary when introducing one organism to control another. They conducted intense testing to make sure the organisms used would be safe and effective. Of the more than a hundred insects that feed on loosestrife in Europe, several beetle species were thought to have excellent potential in North America.

TABLE 6—1:
HERBICIDE USE ACROSS THE STATE ON PURPLE LOOSESTRIFE

YEAR	MONEY SPENT	AMOUNT USED
1989	$102,000	471 gallons
2011	$410	.9 gallon
2016	$0	0 gallons

Source: Minnesota DNR

The beetle Galerucella calmariensis *on loosestrife leaf.* Bill Johnson

dia, Thailand, Malaysia, Burma, and Nepal from the first century CE and beyond.

From about 675 CE until the 1860s, emperors in Japan issued prohibitions on meat eating, in accordance with Buddhist teachings on killing. During that period soybeans became a major portion of the Japanese diet, supplying the protein and savory flavors that had previously come from meat. Not until after World War II did meat eating become a routine part of Japanese culture.

In the sixteenth and seventeenth centuries, European visitors to China and Japan sampled soybean products and noted the dishes in their diaries. They reported on Asian cooks' creativity in converting

Field of Minnesota soybeans. David L. Hansen, University of Minnesota

soybeans into miso, soy sauce, tofu, and soymilk, all products that were unfamiliar in Europe.

Americans were slower than Europeans to become interested in the soybean. The most common method of preparation was to boil whole soybeans, which, as William Shurtleff and Akiko Aoyagi reported in "History of Soy in the United States," sometimes elicited the comment "not desirable for table use." One gentleman from Boston wrote, "They may be delicious to the celestial palate, but my wife found them hard to cook and I found them hard to eat."

The first documented appearance of a soybean crop in North America was near Savannah, Georgia. Samuel Bowen, a former seaman of the East India Company, had visited China and was familiar with the beans and the products that could be made from them. In 1765 Bowen planted soybeans, then known as *Luk Taw*, or "Chinese vetch," near Savannah and produced soy sauce and soybean noodles for the English market. He theorized that the sprouts of the beans had antiscorbutic properties, meaning they could counteract scurvy, and thus would help the British Royal Navy in its fight against the disease aboard its ships. This research led to Bowen receiving the gold medal in 1766 from the Society for the Encouragement of Arts, Manufacturers and Commerce (London) and a gift of 200 pounds from King George III. During 1770–75 Bowen exported 1,058 quarts of soy sauce to England, but the American Revolution greatly reduced any exports and sales. When Bowen died in London in 1779, the soybean enterprise ended.

WHAT MINNESOTANS HAD TO SAY ABOUT SOYBEANS

[Soybean] has affected not only the plant world but the livestock industry as well as the human food chain. In the 1930s from nothing to the second most important crop today in Minnesota.

ROY L. T.

The Soyfoods Association of North America records that in 1770 Benjamin Franklin sent soybean seeds from London to the botanist John Bartram for his garden on the Schuylkill River near Philadelphia. In the letter Franklin described a cheese (tofu) that the Chinese made from soybeans: "My ever dear Friend: I send Chinese Garavances. Cheese [is] made of them, in China, which so excited my curiosity. Some runnings of salt (I suppose runnet) is put into water, when the meal is in it, to turn to curds. These are what the Tau-fu is made of."

American soybean references are sparse for the next seventy-per acre, or 2,898 bushels. In less than ten years, by 1929, US production had grown to 9 million bushels. Through the 1930s, much of the crop was used for silage or as a "plow-down" crop to add nitrogen to the soil.

In the 1920s and 1930s, the University of Minnesota Experiment Stations had access to bean varieties brought from similar latitudes in China and Korea. These were tested at the experiment stations in Waseca and Morris. However, researchers still believed the bean had limited potential. According to the booklet *Food for Life,* the University of Minnesota Experiment Station's 1932 annual report stated, "The soybean crop has an important function in Southern Minnesota agriculture as an annual or emerging hay crop in case of clover hay failure."

By 1937 farmers in southern Minnesota began experimenting with soybeans as a new oil crop. University of Minnesota Extension hesitated to promote the crop until its marketability was certain. Still, farmers persisted in planting the bean, and by the 1940s, Extension held regular soybean variety demonstrations as part of its agricultural programs. In 1940 farmers in southern Minnesota were planting approximately 251,000 acres of beans, with a yield of fifteen bushels per acre.

FACTORS FUELING SOYBEAN PRODUCTION

In 1929 agricultural scientist Walter J. Morse, sometimes called the "father of soybeans in America," began a two-year exploration through Asia, gathering new varieties of soybeans and learning about growing methods and uses for the crop. He returned with 4,451 new varieties, which became the basis for US research and eventual leadership in the field.

In 1932–33 the Ford Motor Company spent more than $1 million on soybean research. By 1935 soy was involved in the manufacture of every Ford car. For example, soybean oil was used to paint the automobile and for shock absorber fluid. Ford's engineers used soy to fabricate a plastic strong enough for gearshift knobs, accelerator pedals, window frames, and other car parts. Henry Ford's involvement with the soybean opened many doors for agriculture and industry to be linked more strongly than ever before.

World War II brought great impetus to soybean growing in the United States. The main supplier at the time, China, was at a standstill due to the war and its own internal revolution. As the United States entered the war, there was a steep increase in demand for oils, lubricants, plastics, and other products. These needs expanded the market for soybeans, spurring farmers to increase production.

A big help for Minnesota came in 1946 when the university hired Jean W. Lambert as a professor to research plant genetics, including soybean varieties tailored to the state. Lambert helped develop Minnesota's first soybean variety, Renville, in 1953. When he retired in 1983, he had contributed to the development of eighteen soybean varieties adapted to various climates and conditions in the state. Lambert's work, which was crucial in making Minnesota one of the nation's top soybean exporters, was honored in an address by Representative Gilbert W. Gutknecht to the Minnesota House of Representatives in 2000.

After the Second World War, the United States experienced a period of increasing prosperity, which created a demand for meat consumption. Livestock farmers in Minnesota and surrounding states responded by expanding their operations. With more chickens, cows, hogs, and turkeys to raise, farmers turned to diets of soy meal because it was a low-cost, high-protein feed ingredient. Crop farmers began converting to soybeans as they became commercially productive. In 2014 the Minnesota livestock industry consumed the meal from 68.2 million bushels of Minnesota soybeans.

At the same time, Minnesota business leaders quickly responded to the needs for processed meal and oil and constructed soybean crushing plants in the state. Having processors close by made it easier and more profitable for farmers to get their harvest to market. The first soybean processing plant in the state, Mankato Soybean Products, Inc., was constructed in Mankato in 1939. Others soon followed: Honeymead and Archer-Daniels-Midland (now ADM) in

During World War II the US government promoted soybeans as a protein replacement and a source of edible oil. Courtesy National Archives

YOUR COUNTRY NEEDS

SOYBEANS

for

★ FOOD

★ FEED

★ GUNS

ROW MORE IN '44

Mankato, Cargill at Port Cargill, and five others in Lakeville, Preston, Blooming Prairie, Glencoe, and Dawson. According to Ray Goldberg in his book on the Minnesota soybean industry, by 1951 the state's processing capacity was approximately 12 million bushels a year. The companies' products included soybean meal, soybean oil, lecithin, crude soybean oil, toasted soy flour, and degermed oil.

Continual innovations in processing enabled producers to efficiently obtain more product from each bushel. The original screw-press and hydraulic processes (essentially mechanical methods) were replaced by solvent extractors, resulting in an additional pound and a half of oil from one bushel. Improvements in the solvent method have continued to evolve.

The Minnesota legislature recognized the importance of soybeans to the state's economy and in 1960 increased funding to expand work in plant genetics and physiology. Two years later in southwestern Minnesota's Sleepy Eye, a handful of farmers got together to establish a nonprofit, farmer-controlled organization that would sponsor research and market development. The Minnesota Soybean Growers Association has as its mission research and marketing to increase farmers' profits, price received for their crops, and production. This group has grown from a small number of members to become a powerful organization with 3,600 farmer members.

In 1973 the Minnesota legislature approved the state soybean checkoff program, which mandates that each farmer is assessed one half of one percent of the net market value, or .5 percent of the market price for every bushel of soybeans sold. The federal legislation permitting a soybean checkoff program on a national level was passed in 1991. The United Soybean Board was developed to disperse the funds collected by the checkoff and is run by a farmer board based on the number of bushels produced. These dollars are divided fifty-fifty between national and state checkoffs and are channeled into research and promotion of soybeans. Over the years, the dollars have funded proposals to educate farmers and the public, programs to promote uses of soy, and research to improve yields, agronomic practices, and pest management.

WHAT MINNESOTANS HAD TO SAY ABOUT SOYBEANS

It amazes me how this plant is used in so many ways. I wasn't kidding. It is an amazing bean that shapes not only agriculture, but manufacturing, the medical field, and many inventions! Most people do not realize how many products use soy beans.

MICHELLE H.

In 1981 Dr. James Orf joined the University of Minnesota staff as a soybean geneticist and breeder, following in the footsteps of Lambert. Orf, who received the Outstanding Achievement Award from the soy checkoff program in 2015, is credited with creating more than fifty general-purpose soybean varieties, as well as more than sixty special-purpose varieties throughout Minnesota.

Prior to earning his PhD, Orf had served for two years in the Peace Corps in Kenya. There he was drawn to the idea of plant breeding and increasing the protein in soybeans to improve people's lives. Much of his work at the university has involved food-grade soybean variety development as well as work on higher yields, earlier maturity (he helped expand the growing range farther north), pest resistance, and niche markets.

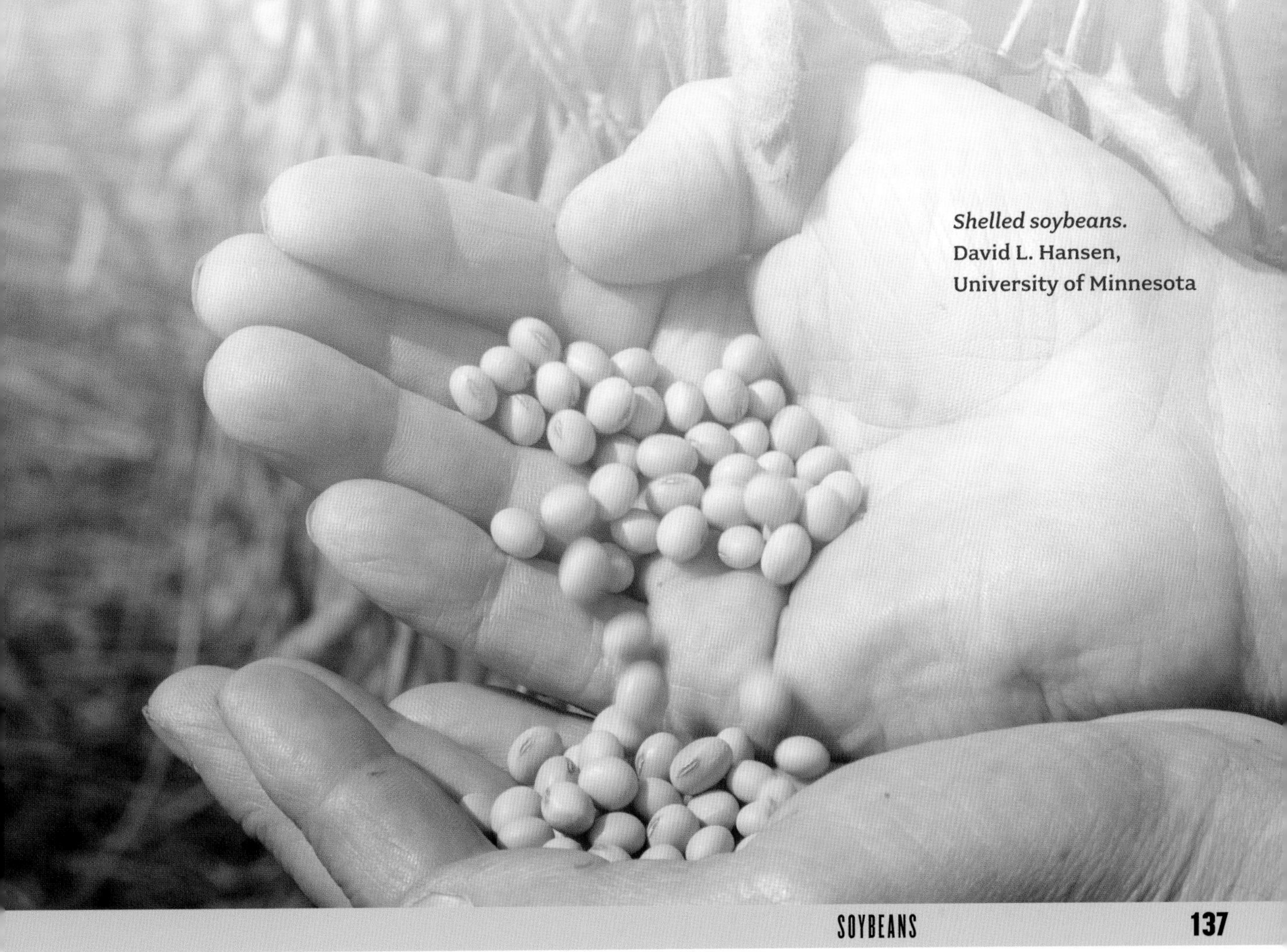

Shelled soybeans. David L. Hansen, University of Minnesota

The breeding efforts by Lambert, Orf, and others at the university have paid off. In 1959 Minnesota farmers were producing nineteen bushels of soybeans per acre. Today it is estimated that the state's farmers (about 29,000 in number) get more than twice that amount—52.9 bushels per acre in 2015 on 7.5 million acres of land.

In 1992 Congress passed the Energy Policy Act to encourage the use of biofuels. During the 1990s the soy checkoff funded research and demonstrations on biodiesel benefits for engine performance, the environment, and human health. In 2011 biodiesel fuel qualified as an advanced biofuel under the Renewable Fuel Standard, a federal program requiring that transportation fuel sold in the United States contain a minimum volume of renewable fuel.

There is a large market for soybeans because they are so versatile. They provide more protein per acre than most other uses of the land. Soybean meal is an inexpensive, high-quality protein for animal feed. As a food for humans, soybeans can be eaten in numerous

Soybeans on the vine. David L. Hansen, University of Minnesota

ways—as flour, milk, infant formula, tofu, and soy sauce, to name a few. In industry, their uses are many. Paint, carpet, fuel, resins, and plastic composites are just a sampling of the products that utilize soybeans.

SOYBEANS AND NUTRITION

Soybeans contain proteins, vitamins, and minerals. Because they provide significant amounts of the essential amino acids, soybeans are considered a complete protein, which makes them useful to vegans and vegetarians. Soybeans are also an excellent source of fiber and phytochemicals. Several large studies have shown that consumption of soy foods is associated with a reduced risk of prostate cancer in men and decreased incidence of death and recurrence of breast cancer in women. Consumption of soy may also reduce the incidence of colon cancer.

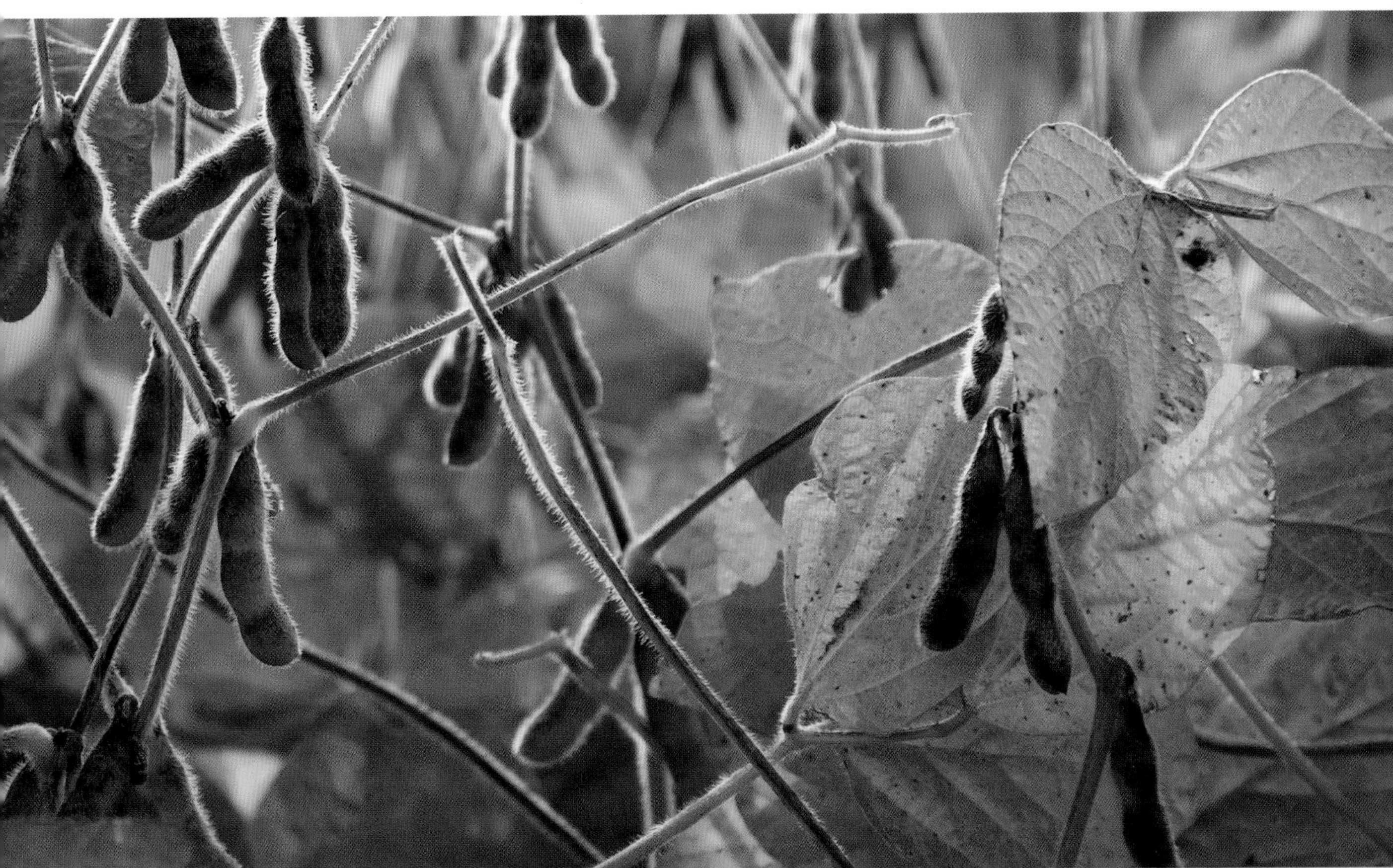

A CONFLUENCE OF OILS

WHEN PEOPLE HEAR "OIL SPILL," they think primarily of petroleum spills. In Minnesota there was a major confluence of soybean and petroleum oil spills in the winter of 1962–63. That year, 1 million gallons of industrial oil from the Richards Oil Company plant of Savage converged with 3.5 million gallons of crude soybean oil from the Honeymead Plant in Mankato, the largest soybean oil processing facility in the world at the time.

On the morning of January 23, 1963, a Honeymead storage tank burst after becoming brittle and contracting in the bitterly cold weather (the temperature that morning measured twenty-five degrees below zero). The explosion sent a thirty-foot-high wall of yellow soybean oil into the air, covering Mankato streets, yards, and garages nearly three feet deep and then cascading into the frozen Minnesota River. The force of escaping oil pushed the big storage tank backward into five smaller tanks holding a total of half a million gallons of salad oil. These burst as well, adding salad oil to the soybean oil. The enormous wave of soybean oil felled hundreds of barrels of lecithin and other products in its path, blew out the doors of the extraction building, and overturned two railroad cars, weighing eighty tons each.

In the cold air, the oil quickly cooled to a gooey, congealed mess. Some residents were trapped in their homes; many had cars and equipment that became surrounded by the pool of oil. Initially, Honeymead employees and city workers exacerbated the problem by using snowplows and graders to push oil off roads into ditches, ensuring its eventual flow into the river. The Minnesota Department of Health quickly halted the practice. In addition, Honeymead employees took oil out in trucks and dumped it into a ravine by the Le Sueur River, not realizing that the area would melt in the spring, spilling oil into the river. Later, Honeymead built dikes along the riverbanks to attempt to contain the oil.

Downriver in Savage, a pipeline had burst from the cold on Friday, December 7, 1962, sending 1 million gallons of oil over thirty acres of ground. Richards Oil Company, a family-owned business, was unstaffed on the weekend, and the spill remained undetected and undiverted until Monday morning. By then, the oil was pouring into the Minnesota River, spreading downstream on top of and below the ice.

Richards Oil Company's owners and employees repeatedly denied any problems at the plant except a minor spill and routine leaks; the owners did not acknowledge the spill until March 18, 1963. By late March the ice had gone out, and most of the oil had moved downriver into the Mississippi. The soy oil flowed downriver, merging with the petroleum oil; this mass moved into the Mississippi River and Lake Pepin, coinciding with the peak of waterfowl migration.

On March 28, as recalled in "Operation Save a Duck," sixteen-year-old John Serbesku noticed "dark blobs struggling in the murky waters" near Pine Bend. The ducks had been caught in the floating oil. Immediately the Serbeskus began a one-family rescue operation, removing oil-covered ducks from the river and washing them in their basement. The family called for volunteers, boats, and governmental action. After initial hesitation, conservation and wildlife officers made intensive duck rescue efforts. They washed more than a thousand ducks in detergent, soaked them with a trisodium phosphate rinse, and housed them until fall, when they molted and grew new feathers. Out of the ducks that were caught, some 350 survived and flew away.

At the time there was little understanding of the environmental catastrophe that was occurring. Conservation officers, plant owners, and citizens were mostly worried about the oil destroying the fish in the spring, not realizing how far downriver it would flow. Most of the fish seemed unharmed; indeed, carp could be seen eating the soy oil. But an estimated 30,000 ducks were killed and other aquatic life destroyed along parts of the Mississippi.

The disaster sparked outrage among citizens who viewed ducks flopping on shore and sinking in the oil. The river pollution caused by these spills brought the condition of all of Minnesota's waters to the fore. Responding to the public's shock and indignation, the legislature took action. On May 22, 1963, Senator Gordon Rosenmeier's bill to protect Minnesota's waters passed in the senate and the house after much maneuvering among lawmakers concerning its components. The final law included requirements for liquid-storage safeguards, a municipal sewage treatment takeover provision, and an enlarged definition of state waters that included underground water. In May 1967, the legislature passed a bill creating the Minnesota Pollution Control Agency, thus providing Minnesota with an oversight group. The oil spills of 1962–63 and the resultant duck deaths had the ultimate effect of awakening Minnesotans to the dangers facing their waters and giving them the resolve to provide protections.

Cleaning up in spring after an unusual soybean oil spill. US Fish and Wildlife Service

SOYBEANS *and the* ENVIRONMENT

SOYBEANS ARE PRODUCED as a monoculture on tilled soil, a practice that can impact the environment, with possible nutrient runoff and soil erosion. Soybeans usually require weed and insect control, and often supplemental fertilizers. Although soybeans, like alfalfa, can fix nitrogen and require much less fertilizer than corn or wheat, they still need careful management to eliminate runoff of fertilizers or pesticides into nearby streams or lakes. As more and more row crops (corn and soybeans primarily) are planted, animal habitat and native vegetation are reduced.

In the last few years, neonicotinoids, insecticides with low human toxicity, have been used to coat soybean seeds before planting. The seed is treated in an effort to thwart pests, such as the soybean aphid, which can reduce and greatly impact soybean yields. However, new research has shown that neonicotinoids can affect honeybees, which inadvertently take the chemicals back to their hives. A recent report by the extension departments of twelve public universities, including the University of Minnesota, concludes that the neonicotinoid coating does not protect against the major soybean pest, the soybean aphid. In addition, neonicotinoids are highly soluble in water and can move beyond the fields on which they are used, thereby contaminating nearby bodies of water.

In an ongoing balancing act, farmers are continually improving their management practices to minimize the environmental impact while enhancing yields as much as possible. If the application of chemicals and the amount of land cultivated are minimized, farmers can continue to provide food for animals and people, while reducing the impact on the environment and protecting our heritage for future generations.

A university soybean field spells out a giant "M."
David L. Hansen, University of Minnesota

DID YOU KNOW?

 The **SCIENTIFIC NAME** for soybeans is *Glycine max*.

 The Chinese grew soybeans at least **5,000 YEARS AGO**.

 The United States consumed **17,700 MILLION POUNDS** of soybean oil in 2011. Soybean oil, the most widely used vegetable oil, is found in margarines, salad dressings, canned foods, sauces, bakery goods, and processed fried foods.

 **REDWOOD COUNTY**, in west central Minnesota, produces the most soybeans in the state.

 One acre of soybeans can produce **82,368 CRAYONS**, enough to fill 3,432 boxes of 24 crayons each.

 **MORE SOYBEANS** are grown in the United States than anywhere else in the world.

 Soybeans are a major ingredient in **LIVESTOCK FEED**.

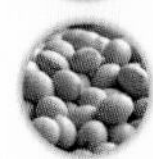 Soy is a **GOOD SOURCE OF PROTEIN** for humans and animals.

 Soybeans **ADD NITROGEN** to the soil.

 Soy oil is used in **INKS, VARNISHES, AND PAINTS**.

INDUSTRIAL *uses for* SOYBEANS

adhesive
agricultural adjuvants
all-purpose lubricants
alternative fuels
analytical reagents
animal care products
antibiotics
anti-corrosion agents
anti-foam agents
  alcohol
  yeast
anti-spattering agents
anti-static agents
asphalt emulsions
auto care products
bar chain oils
binders—wood resin
biodiesel fuel
building products
calf milk replacers
candles
carpet backing
caulking compounds
cleaning products
cleansing materials
composites
concrete supplies
core oils
cosmetics
crayons
dielectric fluids
diesel additives
disinfectants
dispersing agents
  inks
  insecticides
  paints
  rubber
dust control agents
dust suppressants
electrical insulation
engine oils
epoxies
fermentation aids/nutrients
films for packaging
filter material
fuel additives
fuel oil emulsifiers
fungicides
furniture care products
hair care products
hand cleaners
home and lawn products
hydraulic fluids
industrial cleaners
industrial lubricants
industrial proteins
industrial solvents
insulation
leather substitutes
linoleum backing
lubricants
margarine
metal—casting/working
metalworking fluids
odor reduction
oiled fabrics
paint strippers
paints—water based
paper coating
particle boards
personal care products
pesticides
pesticides/fungicides
pharmaceuticals
plasticizers
plastics
polyesters
printing inks
printing supplies
protective coatings
putty
resins
saw guide oils
shortening
soap/shampoo/detergents
solvents
stabilizing agents
textile fibers
textiles
two-cycle engine oils
varnishes
vinyl plastics
wallboard
waterproof cement
waxes
wetting agents

SOURCE
American Soybean Association

Crayons, one of many products that can be made from soybeans. iStock.com/Pamela Moore

EDIBLE uses for SOYBEANS

HUMAN CONSUMPTION
alimentary pastes
antioxidants
baby food
baked soybeans
bakery ingredients
bakery products
baking applications
batters and breading
beer and ale
beverage powders
bread and rolls
breads/specialty
breakfast cereals
cakes and cake mixes
canned meats
cereals
cheeses
coarsely chopped meats
coffee creamers
coffee whiteners
cookies
cooking oils
desserts
dietary supplements
  phytosterols
  soy isoflavones
  vitamin E
doughnuts
emulsified meats
emulsifying agents
  bakery products
  candy/chocolate coatings
  pharmaceuticals
food drinks
frozen dairy desserts
full fat soy flour
  bread
  candy
  doughnut mix
  frozen dessert
  instant milk drinks
  low-cost gruels
  pan grease extender
  pancake flour
  pie crust
  sweet goods
gravies
grits
high-fiber breads
hypoallergenic milk
infant formulas
margarine
mayonnaise
meat products
noodles
Oriental foods
pancakes
pasta products
pastries
peanut butter
pharmaceuticals
prepared mixes
roasted soybeans
  candies/confections
  cookie ingredient/topping
  crackers
  dietary items
  fountain topping
  soy coffee
  soynut butter
salad dressings
salad oils
sandwich spreads
sauces
sausage castings
shortenings
snack foods
soups
soy flour concentrates
soy flour isolates
soy sprouts
sweet rolls
traditional soyfoods
  miso
  soy sauce
  soymilk
  tempeh
  tofu
whipped toppings
whole muscle meats

ANIMAL CONSUMPTION
aquaculture
bee foods
calf milk replacers
cattle feeds
dairy feeds
fish food
fox and mink feeds
milk replacers for young animals
pet foods
poultry feeds
protein concentrates
soybean meal
stock feeds
swine feeds

SOURCE
American Soybean Association

These lists from Soy Stats 2012 show the wide range of products that come from soybeans and how they are used.

8
WHEAT

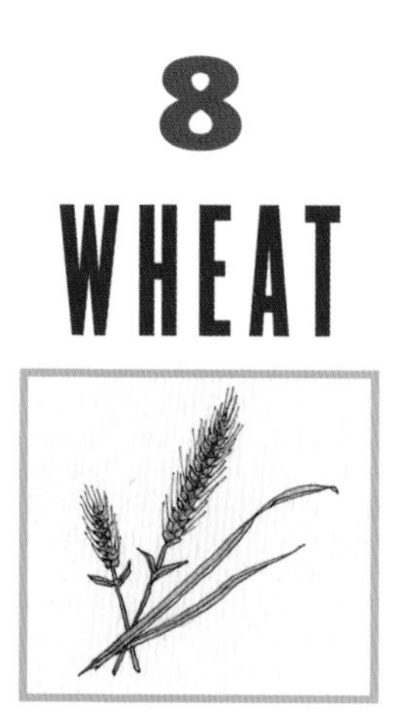

Wheat everywhere; wheat on the levee;
. . . wheat in the streets; wheat in the side-walks;
warehouses of wheat; . . . men talking of wheat.
FROM MERRILL E. JARCHOW, "KING WHEAT"

Wheat is a primeval grain, originally grown in the lands along the eastern Mediterranean known as the Levant. The ancestors of our wheat are the primitive cereals emmer (*Triticum dicoccum*) and einkorn (*T. monococcum*). Historians report that about 17,000 years ago people chewed on the kernels of the wild grains and so discovered their tastiness and nutritive qualities. Through domestication, human beings created a larger grain that would not shatter at harvest.

Wheat was one of the first cereals to be cultivated, and its development contributed to the emergence of city-states, including those in the Babylonian and Assyrian Empires. It could be grown on a large scale, and the harvest could be stored for long periods. The ancient world recognized the importance of wheat, which supplied a significant portion of their protein.

Some of the earliest remains of the crop have been discovered in Syria, Jordan, and Turkey. Primitive relatives of today's wheat have been found in some excavations in Iraq that date back 9,000 years.

Other findings show that bread wheat was grown in the Nile Valley about 5000 BCE. The cultivation of wheat seems to have spread out from the Fertile Crescent, beginning around 8000 BCE, reaching Greece, Cyprus, and India in about 6500 BCE. China was growing the grain by 2500 BCE, as was England by 2000 BCE. Around 6000 BCE, people discovered that the grain could be ground with stones rather than teeth, and thus simple stone milling began.

Many of the world's early religions considered wheat the symbol of life. These ancient peoples glorified the harvest and venerated bread as a gift of the gods. The Egyptians worshipped Isis, the goddess of fertility and agriculture who, according to myth, had discovered wheat and barley. Isis's Greek counterpart was Demeter,

Minnesota wheat field. David L. Hansen, University of Minnesota

WHAT MINNESOTANS HAD TO SAY ABOUT WHEAT

[Wheat] created a way for settlers to earn a living in rural Minnesota. Helped to build an industry that encouraged economic growth in cities along waterways.

JACK S.

the goddess of the harvest, who presides over grains and the fertility of the earth. The Romans had Ceres, credited with discovering spelt wheat and giving the gift of agriculture. The Jews celebrate the abundance of wheat in their festival known as Shavuot.

Wheat was brought to Minnesota by the early Euro-American farmers. Bread was essential to their diet, considered the "staff of life" by most immigrants. Colonel Josiah Snelling sent soldiers to build the territory's first flour mill on the west side of St. Anthony Falls in 1823. Anecdotes of the time relate that the soldiers were poor millers, allowing the flour to mold. Thus the bread was black and tasted bad, nearly causing a mutiny at the fort among the men.

Later wheat was planted sporadically by a few more of the newly arrived Euro-Americans. The first commercial mill was built in 1845–46 by Lemuel Bolles in Afton. According to the 1850 US Census, only 1,401 bushels of wheat were grown in the Minnesota territory, and three millers were listed.

From 1850 to 1855, a dozen small gristmills were constructed in the counties of the territory along the southeastern rivers, the Rum, Cannon, and Minnesota. The mills were to be found in Houston, Rice, Winona, Wabasha, Dakota, Ramsey, Washington, Chisago, Sherburne, Stearns, and Hennepin Counties. Some of the earliest ones, such as those in St. Peter and Mankato, were powered by wind, but most were water-driven.

Much of the wheat used for these mills was hard spring wheat, a variety known as Scotch Fife, imported from Illinois and other areas south of the state. Winter wheat (sown in the fall and harvested in the spring) was not as reliably hardy here. Spring wheat (planted in the spring and harvested the same year) avoided the winter but was difficult to process with the equipment early millers owned. Theirs was traditional technology that pulverized the grain and ground as much flour as possible in one pass. This method created a product that was disliked by consumers because it had a gray, speckled appearance and contained bits of bran.

In the 1860s several men began to experiment with the high grinding of wheat, a process used in Canada on hard wheat. In this method, the millstones were set slightly farther apart than normal, causing the kernels to break and crack rather than being instantly pulverized. The wheat was extracted at slow speeds and removed gradually to avoid overheating or damaging it. This method, which resulted in whiter and purer flour, was called the "New Process." In addition, these millers used a light blast of air to separate the bran from the wheat. Early on, Stephen Gardner of Hastings, brothers John S. and E. T. Archibald of Dundas, A. G. Mowbray of Winona, and John W. North and Alexander Faribault of Northfield were the principal producers of the finely ground white flour. They kept their methods secret, in hopes of maintaining a corner on the market. It is not certain which man was the first to use "high grinding," but it is thought that Gardner was the innovator.

Where these millers had learned about the method is not certain either. It was used in Canada by French millers, and several of the Minnesotans had learned their craft there and perhaps brought the method to southeastern Minnesota. Another theory holds that Faribault, whose father was a French fur trader, was aware of the French method of producing fine white flour. Faribault brought to his mill two highly educated millwrights from France, the brothers Nicholas and Edmund N. LaCroix. They not only set the grinding stones high but also built various purifiers to remove the bran and impurities. They worked for Faribault, the Archibalds, and probably others, and their product proved that the French process could be used to mill the hard, flinty, and cheaper wheats of the northern Midwest.

Because of its uniformity and baking characteristics, the finer white New Process flour did extremely well at market and soon became the envy of other millers. Accounts of the time relate that the Minneapolis millers were extremely curious about the flour and eager to discover the reasons for the exceptional product. Charles A. Pillsbury made a probing expedition to the Archibald Mill on the Cannon to uncover the secret of the superior quality obtained there. Thinking that the high quality must be the result of the wheat being used, Pillsbury managed to "pocket" a handful

The Pickwick Mill on Big Trout Creek near Winona, built in 1856. Still open to the public May through October. MNHS collections

of wheat from the Archibald Mill. Returning home, he discovered to his disappointment that the grain was no better than the wheat he was milling.

George H. Christian, a flour broker for the town of La Crosse, Wisconsin, was eventually successful in ferreting out the secret process. Patiently, he visited the Dundas mill again and again. On each visit, Archibald would boastfully mention a point or two about his methods, never realizing that Christian was keeping notes on his comments. Approaching Cadwallader C. Washburn in Minneapolis, Christian explained what he had learned. Then, in the early 1870s, he was hired to manage the Washburn B Mill. In this way the French processes were introduced to the Minneapolis mills. Soon Christian hired Edmund LaCroix to build "middlings" purifiers and experiment further in the perfection of the products. In this method, a vibrating sieve fed the ground wheat into a purifier, which used suction and air jets to remove the light bran. Heavier impurities remained on the sieves, separating the bran from the usable part of the grain. The resulting flour was the basis for Washburn's enormous success.

New Process flour quickly became a thoroughly established practice, and mill owners upgraded their equipment to meet the standard. This method of gradual reduction and middlings purification was used in the North Star State earlier than in any other, giving Minnesota an advantage. Not only American markets but foreign outlets were important from the start. For several decades one-third of the Minnesota flour product was shipped abroad.

"KING" WHEAT

Minnesota-grown wheat became commercially important in the late 1850s. In the first years of the territory, wheat had to be shipped upriver from Iowa to St. Paul, and from there it was hauled by horse teams to the mills. This caused Horace Greeley, editor of the *New York Tribune*, to caustically write in 1858, as John Storck and Walter Dorwin Teague included in their book *Flour for Man's Bread*, "I saw that your state [Minnesota] imported not only loafers in great abundance, but the bread they ate as well as the whiskey they drank."

Once the Euro-Americans started growing wheat, the many southeastern rivers and St. Anthony Falls provided the power, and the Mississippi River and later the trains provided the transportation. By 1860 the total state wheat production was 2,186,973 bushels, impressive compared with the total of only 1,401 bushels a decade earlier. Also in 1860 exports of 1,576,666 bushels were shipped to St. Louis, Missouri, and Lockport, New York.

In 1873 Red Wing became the world's largest primary wheat market, with 1.8 million bushels of grain exported and another 950,644 in storage. This activity drew investors—bankers, middlemen, lawyers, and various entrepreneurs—to partake of the bounty of wheat. The *Illustrated Atlas of Minnesota* in 1874 gave Goodhue County the label "the banner wheat county of America." Unfortunately, by 1878 stem rust began to infest many crops in Goodhue County and other southeastern areas. As this happened, local farmers diversified their crops, and major wheat growing moved westward to other parts of the state.

In 1880 almost 70 percent of Minnesota's farmland (4.5 million acres) was planted in wheat, producing more than 34 million bushels of wheat that year. The state's weather, particularly its cool nights, allowed wheat to develop more protein, making the wheat grown here a commodity of high value.

As was true for all frontier areas, the main limitations in moving wheat were marketing and transportation options. According to the 1861 state legislative report, the mean distance for a Minnesota farmer from his field to a navigable river was eighty miles. With his thirty-bushel wagonload, the farmer made a six-day journey there and back. The roads were rough and the temperatures often low. Farmers might be attacked by robbers; some even froze to death. The river towns often dealt with congestion on the streams and ice in winter.

Much of wheat growing was a gamble. Periods of drought caused crop losses. Fungal diseases such as stem rust and smut struck frequently. To combat this, farmers created a number of potions. In one "remedy," six bushels of seed grain were mixed with three pounds of sugar and chimney soot. In another, one ounce of blue vitriol, or copper sulfate, was dissolved in a pint of water and poured on each bushel of wheat. Interestingly, modern research shows that blue vitriol is effective in controlling wheat smut. When one wheat variety suffered losses, farmers switched to others. Golden Drop, China Tea, Fife, and Russian Red Osaka were planted. The rotations helped, but the grain continued to have problems. In addition, wheat was a heavy feeder and used much of the soil's nutrients, so over time plots would stop producing.

One of the most disastrous problems was the invasion of grasshoppers (Rocky Mountain locusts [*Melanoplus spretus*], now an extinct species), which caused extensive destruction in 1856, 1857, 1864, and 1865. The plague years of 1873–77 were the worst, with

Joseph Askew threshing on Hubbard Prairie, late 1800s. Courtesy of the Wadena County Historical Society

500,000 acres in forty counties badly damaged. The insects arrived in swarms and blocked out the sun. They ate crops to the ground, the wooden handles of implements, and clothing left on the line.

Farmers did their best to destroy the grasshopper eggs. They dragged heavy ropes through their fields to crush the insects. They raised chickens and birds to eat the insects and used flails to beat the grasshoppers. Farm women and children threw blankets over family gardens; the locusts devoured the blankets and then ate the plants. Farmers raked the locusts into piles and burned them like leaves. In the last years of the infestations, they dragged "hopper dozers," sheets of metal covered with molasses or coal tar, across the fields. None of these methods was successful. But in the summer of 1877, the grasshoppers vanished as speedily as they had come. A fortunate April snowstorm had damaged many of their eggs, and farmers dealt valiantly with the remainder. Laura Ingalls Wilder described the episode in her book *On the Banks of Plum Creek*. The novel *Giants in the Earth: A Saga of the Prairie*, by Minnesota author O. E. Rolvaag, spoke of the locusts as one of the many hardships faced by families living on the prairie.

Over time, there were improvements in transporting the wheat. Middlemen in small towns began to buy the wheat in quantity, relieving the farmers of a long haul. In turn, the middlemen would sell the wheat to large mills or dealers. Also, railroads spread west and north across Minnesota, allowing wheat to move quickly. Grain elevators were built alongside the tracks, providing storage points. When farmers were successful, wheat was a fine cash crop. Land was available and cheap in much of Minnesota, and wheat farming required little initial cash outlay. In a good year, wheat was a smart investment. Minnesota ranked first in wheat production in the United States in 1889 and again a decade later. Production had nearly tripled in that ten-year span.

BONANZA FARMS

BONANZA FARMS were enormous farms, ranging from 3,000 to 30,000 acres, that came into being as a result of the failed financial investments of the Northern Pacific Railroad in the 1870s. To pay back investors, mostly easterners, the bankrupt company allowed them to exchange their bonds for land. These men created bonanza farms in the Red River Valley of western Minnesota and eastern North Dakota, making the region one of the country's largest wheat-producing areas. The absentee owners hired local managers to run the farms, using cheap hired labor and highly mechanized equipment. Between 1875 and 1890, these huge acreages were enormous moneymakers. Over time, however, the land became exhausted, the farms were no long profitable, and the investors sold out and moved on. By the 1920s, the bonanza boom had busted.

Oliver Dalrymple's huge bonanza farm in the Red River Valley, the last area of the state where wheat was planted. Institute for Regional Studies, North Dakota State University, Fargo

MILLING INDUSTRY GROWTH

Keeping pace with wheat harvests was the rapid rise of flour milling. At first mills were built wherever water could be harnessed to run a mill. But increasingly they were built in Minneapolis along the Mississippi River at St. Anthony Falls, which provided waterpower. Railroads began linking Minneapolis to the West in the 1860s; the Great Northern Railroad connected Duluth to North Dakota and points west in the 1870s.

By 1876 there were eighteen flour mills along the Mississippi in Minneapolis. Some millers, like Cadwallader Washburn, looked abroad for better engineering processes. Washburn sent his engineer William de la Barre to Europe to unearth the secret methods of Hungarian millers (clearly, those working in the milling industry often kept advanced technology to themselves). Disguising himself to avoid detection, de la Barre took notes on the engineering design and equipment he observed in Hungary. After he returned to Washburn's mills, he installed steel rollers to replace the millstones. This equipment was easier to maintain, prevented heat discoloration, and minimized the crushing of the bran husk. The rollers broke down the grain gradually, integrating the gluten with the starch, and the change to steel meant that the flour was no longer contaminated with the fine grit of the sandstone millstones. In addition, more flour was produced from a pound of grain, and the product had a longer shelf life because the bran was removed. Soon other millers installed rollers, not grindstones, and used the slower grinding method.

The flour industry grew steadily but with one major setback. On May 2, 1878, the four-year-old Washburn A Mill, located on the west side of the Mississippi, exploded when grain dust caught fire. Three massive explosions killed eighteen workers and destroyed the mill. The explosion thrust debris hundreds of feet in the air and could be heard ten miles away. Within minutes, the fire spread to nearby mills, killing four other workers and destroying buildings and residences.

Immediately, Washburn announced that he would rebuild and would install the newest equipment. He stated, as Theodore R. Hazen's "Flour Milling in America" records, "In the new structure I shall adopt the Hungarian system of gradual reduction using rollers, and shall dispense with a great amount of machinery heretofore used, substituting hand labor, which is safer, and I am not sure it is also more economical. I propose to build for an abundance of room and thorough ventilation."

By 1880 Washburn's new A Mill was in operation, equipped with the best technology. The mill had steel rollers, gradual reduction practices, and a safer ventilation system, the Berhns Millstone Exhaust System. The Berhns system greatly reduced the amount of flour dust in the mill, making explosions and fire less likely. With innovations in place, the mill could utilize spring wheat, which was more nutritious than winter wheat and yielded 12 percent more flour. Millers and consumers recognized this wheat for its high quality and attractive appearance.

These developments spawned a significant increase in mill construction. Between 1870 and 1880, seventeen more large mills were built in Minneapolis. In addition, twelve mills opened in the Duluth-Superior region between 1880 and 1898, making use of the Twin Ports shipping area and the railroads. Duluth's growth was fueled by the large amount of grain arriving from the Red River Valley. For a time, Duluth's six-story Imperial Mill was the largest flour mill in the world, producing 6,300 barrels a day. Trains brought the Red River wheat; hours later, flour was loaded into vessels docked at slip 1.

After five years of secret planning, the Pillsbury Corporation, headed by Charles A. Pillsbury and his uncle, John S. Pillsbury, completed its enormous A Mill on the east side of the Mississippi in 1880. This mill, designed by local architect LeRoy S. Buffington, was not only large and technically advanced but also attractive. When it began operating, the A Mill had an output of 4,000 barrels of flour a day; by 1905, the output had tripled, supporting the mill's claim to being the largest flour mill in the world.

WHAT MINNESOTANS HAD TO SAY ABOUT WHEAT

Minneapolis would not have been established and prospered if it wasn't for wheat being grown in the surrounding areas.

DAN P.

Duluth Imperial Mill advertisement. MNHS collections

In its peak years, the Washburn A Mill processed enough wheat each day to fill more than a hundred railroad cars—an amount that could yield 12 million loaves of bread daily. By 1882 Minnesota emerged as the world's leading flour-milling center, with Minneapolis proclaimed the "Flour Milling Capital of the World" and nicknamed "Mill City." Minneapolis went from being a backwater town to a nationally recognized one. An article by Eugene Smalley in the May 1886 issue of *Century Magazine* noted that Minneapolis was no longer "an obscure village [but] a handsome, busy, energetic town chiefly [because of] its flour-mills."

Between 1880 and 1900, numerous mill mergers took place so that by the early 1900s only three Minneapolis corporations controlled 97 percent of the nation's flour production. They were the Pillsbury-Washburn Flour Mills Company, which became Pillsbury Flour Mills Company; Washburn and John Crosby's Washburn-Crosby Company, which became General Mills; and the Northwestern Consolidated Milling Company, later the Standard Milling Company. For fifty years, from 1880 to 1930, Minnesota was the leading state in flour production, with Minneapolis the largest milling city. In 1915–16, flour production hit its peak at 20.443 million US dry barrels annually.

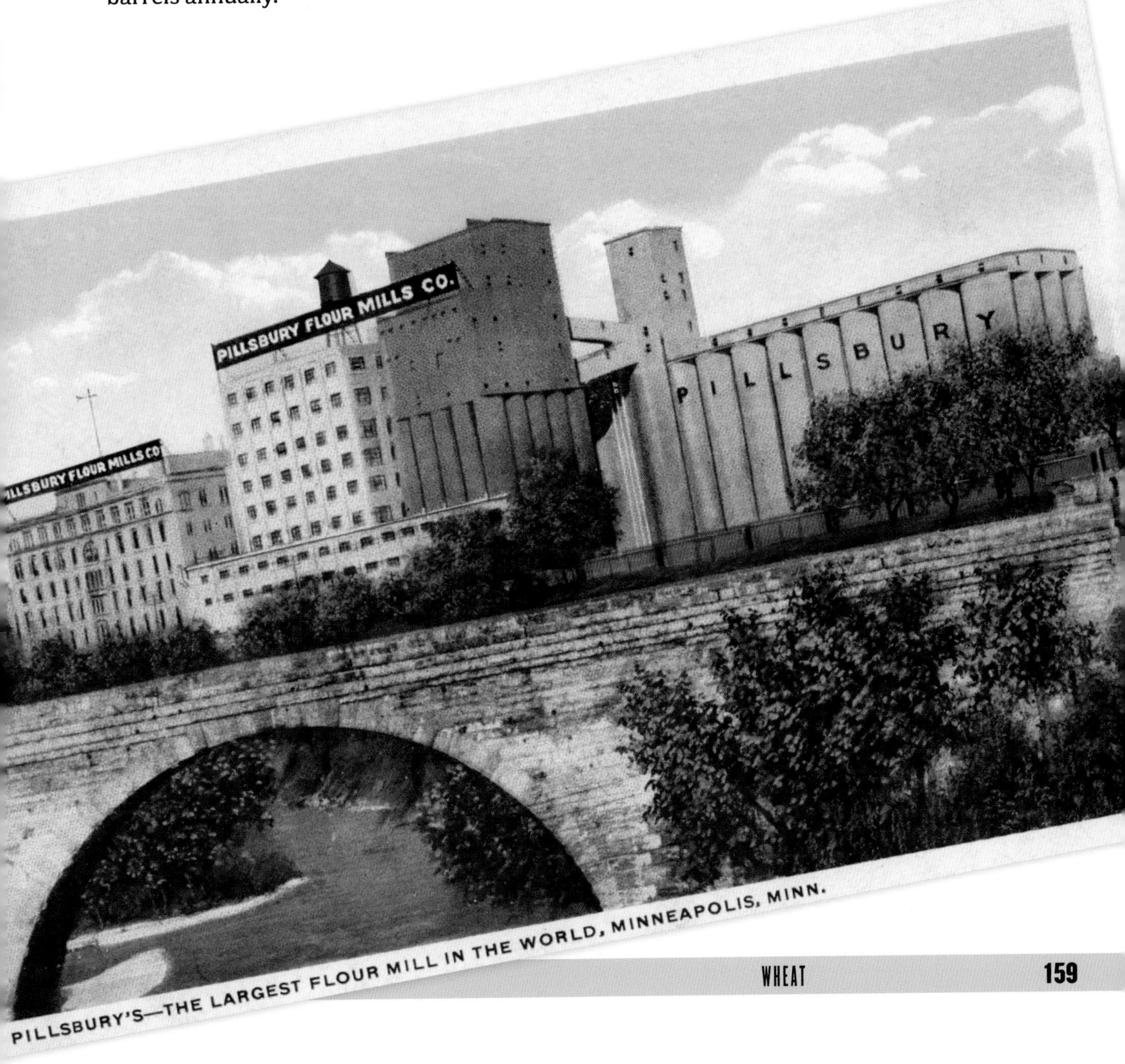

A postcard of the Pillsbury A Mill. MNHS collections

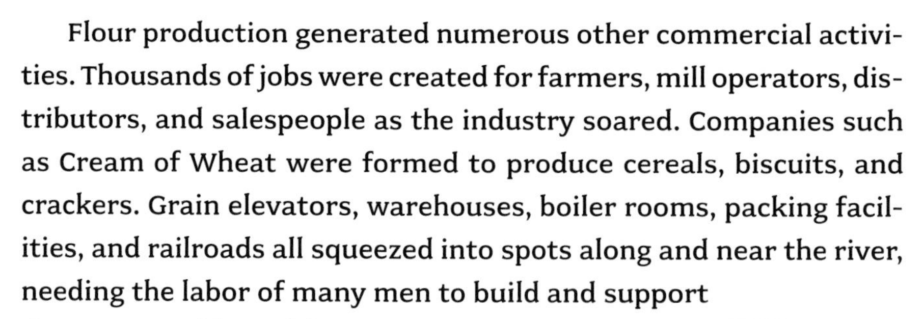

Flour production generated numerous other commercial activities. Thousands of jobs were created for farmers, mill operators, distributors, and salespeople as the industry soared. Companies such as Cream of Wheat were formed to produce cereals, biscuits, and crackers. Grain elevators, warehouses, boiler rooms, packing facilities, and railroads all squeezed into spots along and near the river, needing the labor of many men to build and support them. Bag and barrel factories sprang up to supply the mills. Because of the many mill, lumbering, and railroad accidents, several manufacturers of artificial limbs established their companies in Minnesota. In 1918 Minneapolis was hailed as the leading manufacturer of artificial limbs in the United States.

Milling spurred the growth of railroads, with the Twin Cities becoming a hub due primarily to the enor-

Early advertising postcard Miss Minne Sota—a woman made of wheat. MNHS collections

David L. Hansen, University of Minnesota

mous amounts of grain and flour that needed to be transported. Because milling required large capital investments for plants, machines, and ongoing operations, numerous banks, including Northwestern Bank, National Hennepin Savings, First National Bank, and Farmers and Mechanics Savings, were started to finance the businesses. The extensive banking activity helped establish Minneapolis as the headquarters of the ninth district of the Federal Reserve Bank and contributed to its reputation as the financial center of the Upper Midwest.

In 1881 the Minneapolis Grain Exchange, formerly called the Minneapolis Chamber of Commerce, was created to promote the fair trade of wheat, oats, and corn between the millers and the growers. Often referred to today by the acronym MGEX, it has remained essential to grain trade and is the nation's last independent agricultural futures market. Currently wheat, oats, durum, rye, barley, corn, millet, soybeans, and apple juice concentrate are all traded there.

WHAT MINNESOTANS HAD TO SAY ABOUT WHEAT

Wheat is what grew the city of Minneapolis to become the world leader in flour milling, known as Mill City.

KATE D.

LATER DEVELOPMENTS

After 1900 Minnesota millers began to feel competition from other states. Just as Minnesota had copied European milling processes, millers in other parts of the country began to adopt Minnesota methods. In addition, declining crop fertility and disease problems reduced the quantity and quality of wheat available to local millers. Kansas emerged as a leading wheat-growing state during this time. After World War I, when millers no longer relied on waterpower (here mainly St. Anthony Falls), Minneapolis began to lose its primary place, and other cities caught up to the Mill City in flour production.

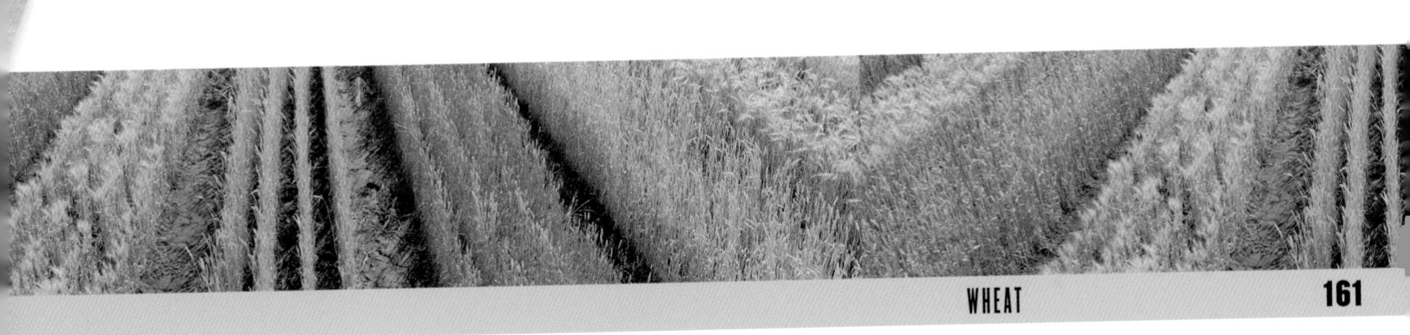

Importantly, Minnesota held on to its customers with new marketing strategies. The Washburn-Crosby Company created the "Gold Medal" label for its flour; Pillsbury named its product "Pillsbury's Best." The Washburn-Crosby Company (which became General Mills in 1928) also developed WCCO radio, using the company initials in the station's call letters. In 1921 Washburn-Crosby created the character of Betty Crocker, who was portrayed as a kindly, knowledgeable homemaker who offered cooking advice. The company developed recipes and produced cookbooks. In 1949 Pillsbury initiated the Pillsbury Bake-Off, a nationwide contest for recipes using Pillsbury products. Companies created other products, such as cake mixes, crackers, and cereals, to replace the sale of flour.

Minnesota wheat growing struggled from 1930 to 1970 due to an aggressive stem rust. This deadly fungus attacked the stems of wheat and spread quickly. In 1935, 50 percent of the combined wheat crop of Minnesota, North Dakota, and South Dakota was lost to stem rust; in 1954, 30 percent of the crop was lost. It was not

SPRING WHEAT VERSUS WINTER WHEAT

THE PRIMARY DIFFERENCE between the two types of wheat is the time of planting. Spring wheat is planted in April or May and is harvested in August or early September. Winter wheat is planted in the fall, starts to grow, then goes dormant during the cold winter months. When the weather warms, the plants resume growing and are harvested in July or August.

In Minnesota, the Dakotas, Montana, and Canada, spring wheat has generally been the chosen crop. Winter wheat is planted less often because of the chance of winter kill. When the winter is particularly severe and the snow cover minimal, farmers can lose their crop. Newer varieties and production practices have reduced this risk, however, and farmers now find that winter wheat is a more viable option than it was in the past. There are benefits to growing winter wheat: the yield potential is higher than with spring wheat; it creates a cover crop to reduce wind and water erosion; it utilizes labor and machinery more efficiently because it is planted and harvested when there is little other field activity; and it establishes a cover for wildlife in fall and early spring.

until 1965 that Chris wheat, a variety that is resistant to leaf and stem rust, was bred at the University of Minnesota. Five years later, Era wheat, the first semidwarf variety, was introduced, spurring a 25 percent increase in crop yield and the worldwide "Green Revolution" in food production.

Research in wheat breeding at the University of Minnesota had begun in 1889, with breeders and a cereal chemist working together to evaluate varieties from Minnesota, Canada, and Russia, Hungary, and other parts of Europe. At the end of ten years, they published a report, *Wheat Facts*, summarizing their research with 552 varieties planted in small plots on the St. Paul campus, noting: "Plant breeding is in its infancy, and plans for extensive scientific breeding of this crop had to be devised rather than copied. . . . Not only yield but the quality of the grain and other characteristics were taken into account in selecting plants to become the mother of varieties."

The University of Minnesota released its first wheat variety, Preston, to farmers in 1895. Since that time, breeders have continued to

OUT OF THE ASHES

FIRES WERE a perennial concern for the mills. The Washburn A Mill suffered two. The first, in 1878, destroyed Cadwallader Washburn's four-year-old building. When rebuilt, the mill was the largest and used the latest technology. At its peak, it was said to grind enough flour to make 12 million loaves of bread in one day. By 1965, though, it was obsolete and essentially abandoned. The second fire occurred in 1991 when the building was home to several tenants and many homeless people. After the fire, the building was in ruins, but thoughtful people realized that the mill still had a story to tell. With help from the Minnesota Historical Society, the ruins were stabilized, much of the structure was saved, and the Mill City Museum was created, opening in 2001. The museum was designed by Thomas Meyer, principal of the Minneapolis architectural firm of Meyer, Scherer & Rockcastle, Ltd. Meyer included much of the original building in his design and used the limestone, brick, steel, and concrete to reflect its industrial origins. The interactive museum tells the story of Minneapolis milling, with exhibits, a movie, a baking lab, and an eight-story elevator ride through time.

study wheat and develop improved varieties. Beginning in 1907, the university's wheat breeding program has been a cooperative project with the US Department of Agriculture—Agriculture Research Service, with plots at the Morris campus. In the intervening years the university has released thirty-eight wheat varieties. Research techniques have changed over the decades, but goals remain the same. Now researchers can make selections for some traits at the genetic level, using DNA markers, rather than selecting chance variations out of thousands of plant crosses. Still, breeders continue to seek high yield, disease resistance, good baking qualities, and the plant's ability to remain upright when harvested.

Today the newest variety of wheat is a hard red spring wheat named Bolles, in remembrance of Lemuel Bolles's 1843–47 mill in Afton, the first in Washington County. Minnesota now ranks third in production of spring wheat, behind North Dakota and Montana. The University of Minnesota reports that farmers recently have expressed a renewed interest in growing wheat and other small grains. According to the USDA National Agricultural Statistics Service, in 2014 wheat was grown on 1.212 million acres in Minnesota, with 66.468 million bushels produced.

HERITAGE AND LANDRACE WHEATS

Heritage wheats are varieties that grew before the "Green Revolution," which started in the 1950s. Modern wheat has short stems; it grows 2.5 feet tall, has a large seed head, and shares genes with dwarf wheat from Japan. The stem heights are even, making harvesting easier. These varieties have a higher yield but generally require greater fertilizer inputs than older varieties. Heritage wheat is much taller (3 to 4 feet) and has a bigger root system.

Landrace wheats are domesticated, traditional varieties that are locally adapted (the word "landrace" comes from the German

Norman Borlaug loved nothing more than working in a wheat field.
David L. Hansen, University of Minnesota

NORMAN ERNEST BORLAUG

NORMAN ERNEST BORLAUG, an American biologist and humanitarian, was born March 25, 1914, on a farm in Iowa. He enrolled in the University of Minnesota, where he studied forestry, receiving a bachelor of science degree in 1937. He received a master's degree in plant pathology and genetics two years later, and earned his PhD in 1942. To pay for his education, Borlaug took various jobs, including one during the Depression with the Civilian Conservation Corps. Here he worked with the unemployed, many of whom were starving, an experience that left a permanent impression on him. In a later interview in the *Dallas Observer* he explained his change of heart: "I saw how food changed them. . . . All this left scars on me."

During World War II he worked for DuPont, an American chemical company, researching compounds useful to the US armed forces. These included camouflage, canteen disinfectants, and saltwater-resistant glue.

Borlaug moved to Mexico City after the war to direct the newly established Cooperative Wheat Research and Production Program, a joint undertaking of the Rockefeller Foundation and the Mexican government. The group of Mexican and US scientists focused on wheat and maize production, plant pathology, and soil improvement. Here Borlaug developed a semidwarf wheat that was high yielding and rust resistant. His work led to increased production of wheat in many countries, including Mexico, Pakistan, and India, and thus improved these countries' food security. Combined, his work and improved agricultural techniques have been labeled the "Green Revolution." Borlaug was credited with saving more than a billion people from starvation and was awarded the Nobel Peace Prize in 1970. He continued promoting these ideas, asserting that increased crop yields would help curb deforestation.

Despite his many accomplishments, Borlaug's work has garnered criticism from environmentalists and nutritionists. The concerns have been on several fronts: for bringing large-scale monoculture farming to countries that had previously relied on subsistence farming, for decreasing biodiversity, for introducing inorganic fertilizers and pesticides, and for promoting genetic crossbreeding. Others lamented the fact that his methods often brought huge profits for US agribusiness and agrochemical companies while also widening the social inequality in countries where they were implemented.

Borlaug listened to his critics but dismissed many of them as elitists. In an interview with the *Atlantic* he maintained that many of those who criticized his approach "had never experienced the physical sensation of hunger. . . . If they lived just one month amid the misery of the developing world . . . they'd be crying for tractors and fertilizer and irrigation canals."

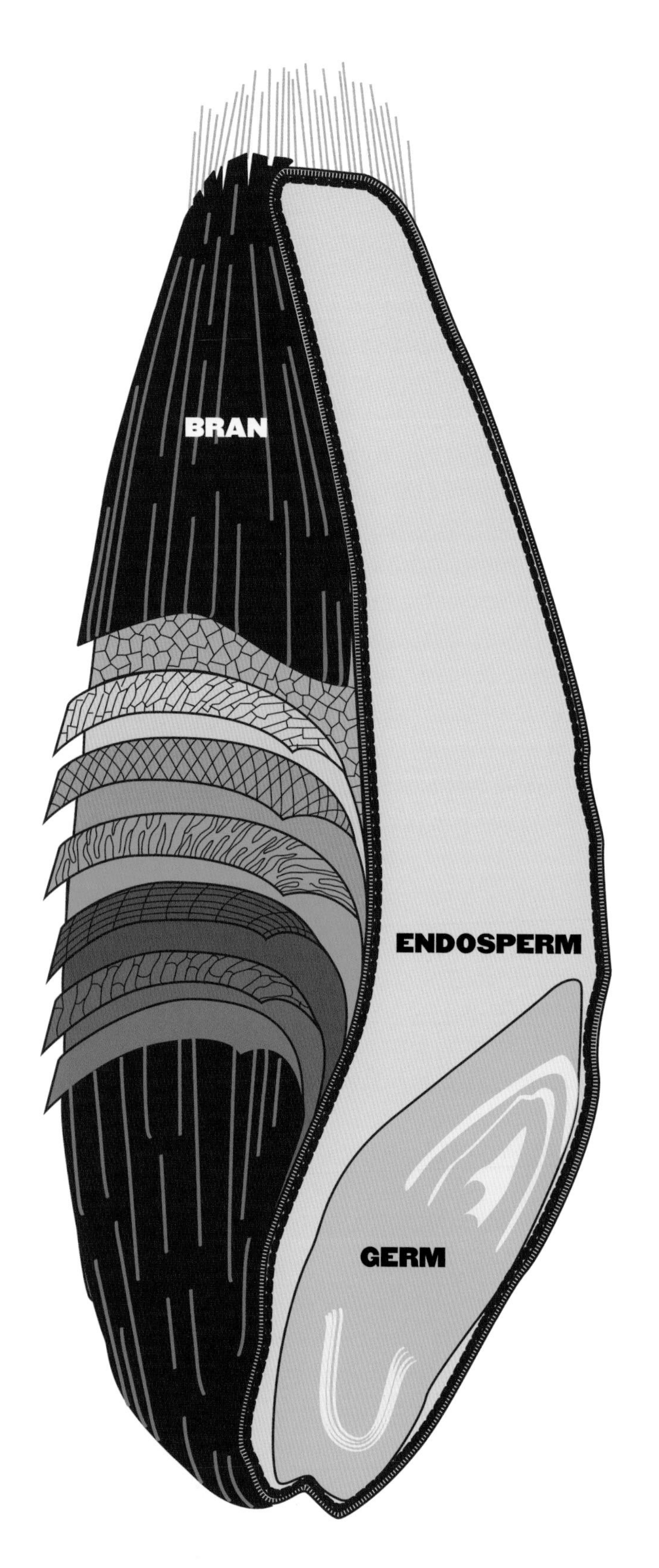

LONGITUDINAL VIEW
OF A GRAIN OF WHEAT

BRAN

The outermost part of the kernel. It contains most of the fiber and is a good source of the B vitamins.

ENDOSPERM

Is made up of mainly starch (carbohydrates). Contains most of the protein, some vitamins and minerals.

GERM

The embryo or sprouting section of the seed. Source of vegetable oils and a rich source of vitamin E and complex B-vitamins.

Graphic by Jennifer Osborne Anderson

A boom site during a log drive. MNHS collections

could fell a few dozen trees a day. To house the cutting crew and the animals, logging camps were built near the cutting sites.

During the spring ice-out, the logs were floated downstream to a boom site. A jam crew worked to ensure that the river did not become clogged with too many logs. Any jam would bring the log drive to a standstill. When that happened, men jumped into the water and onto the logs to remove any obstacles. Log booms like those that had first been used in Maine became standard in the Midwest. Logs were chained across a river, creating a fence to capture the floating logs. Each company had its own crib, a structure to hold the logs in place, with a crew stationed there. Using pikes, the men sorted the logs into the correct crib.

At the mills, the logs originally were cut with a single, heavy iron saw blade, nicknamed a "muley." The muley could slice about 5,000 feet a day when everything was operating smoothly. When several blades were hooked together, in an arrangement called a gang saw,

WHAT MINNESOTANS HAD TO SAY ABOUT WHITE PINE

It started the engine of commerce.

LISA

production was much speedier because a log could be reduced to boards in one quick step. The circular saw, which was introduced in the 1870s, could cut about 4,000 feet an hour.

By the 1860s steam-powered sawmills replaced water mills. With this improvement, timber could be cut both faster and year-round. In addition, the abundance of shavings and sawdust could be fed directly into the furnaces that power the mill. Mills no longer needed to be on a water source and so moved into the northern areas of the state, where uncut trees remained.

With steam power, efficient saws, and draft horses to replace slow-moving oxen, pine logs could be transported ever faster from the forest to the sawmills. As wood was depleted in one area, lumbermen relocated to more plentiful lands around the state. By 1870 there were more than 200 lumber mills in Minnesota, employing thousands of workers. As the sawmills flourished, they spawned ancillary industries—shingle factories, box producers, furniture makers, and window, door, and flooring manufacturers—all needing workers and adding to Minnesota's economic output.

In 1886 the first logging railroads were built in Minnesota. Over the next thirty years, forty different railroads accessed pine stands. In total, 5,000 miles of logging railroads were constructed across northern Minnesota. The railroads not only opened up new areas of timber but also provided a way to distribute finished products. As the railroads replaced river transport, production sped up again. By 1899 Minneapolis was the leading lumber producer in the world, its output nearly 600 million board feet that year.

When Minnesota was at the peak of lumber production (1900–1910), there were 20,000 loggers, 20,000 workers in the mills, and 20,000 more in wood production factories. To skid the logs to railroad spurs and landings, 10,000 draft horses were put into service. The state's yield in 1900 was 2.3 billion board feet of lumber, enough to construct a nine-foot-wide boardwalk to encircle the earth at the equator. For ten years the production continued at the same rate. During the period from 1890 to 1910, the lumber companies' harvest in Minnesota was valued at $1 billion.

When the Ojibwe ceded land to the US government, they were allotted reservation land in return. Two enormous plots in northwestern Minnesota were the Red Lake Reservation, on about 3.5 million acres, and White Earth, established on 800,000 acres. The fine timber of White Earth was estimated at 500 million feet of good, marketable pine. As the acreage available to lumbermen in other parts of the state diminished, they looked hungrily at the reservations. Euro-Americans pouring into the Red River Valley eyed the arable land of White Earth and wanted it for their own. Breaking treaty promises, the United States began to encroach on

A logjam in the Dalles of the St. Croix River, 1886. MNHS collections

FREDERICK E. WEYERHAEUSER

IN THE 1890S the lumber industry, which had started as a scattered grouping of small units, began to concentrate in the hands of a few individuals. Of these, one man stands out: Frederick E. Weyerhaeuser. Born in Germany in 1834, he came to America penniless at age eighteen and in the 1850s moved to Illinois, where he took work at a sawmill in Coal Valley. When the business suffered financial trouble, Weyerhaeuser took over the mill and got it going successfully again. With his brother-in-law Frederick Denkmann, he founded the Weyerhaeuser-Denkmann Company and began purchasing interest in other lumber companies. By 1887 Weyerhaeuser-Denkmann owned many timber sites and mills in Wisconsin and Minnesota.

Weyerhaeuser was a hands-on owner who trekked through huge stretches of Minnesota to make deals—big and small—for timber acreage. In a rare interview he related the reason for his success: "The secret lay simply in my willingness to work. I never watched the clock and never stopped before I had finished what I was working on."

Recognizing that processing of the timber could go more smoothly if he managed the timber transit from stump to mill, Weyerhaeuser established the Mississippi River Boom and Logging Company in 1872, which handled all the logs processed on the river. He continued to buy up small lumber mills, tracts of land, and transport units. In 1891 he moved his family and business to St. Paul, living next door to railroad baron James J. Hill. By 1902 Weyerhaeuser was president of twenty-one different companies. Realizing that midwestern lumber was not inexhaustible, he purchased an enormous tract (900,000 acres) in Washington State from Hill. The incredible deal of five dollars an acre included hugely discounted rail freight rates.

Weyerhaeuser's holdings were so large that in 1905 President Theodore Roosevelt called him "that man whose idea of developing the country is to cut every stick of timber off of it and leave it a barren desert." In fact, by that time Weyerhaeuser understood the necessity for sustainable forest management and had begun the practice of replanting cutover areas. Still later he developed high-yield forests that increased timber production while requiring less acreage.

At the time of his death in 1914, Weyerhaeuser's timber holdings spanned millions of acres of American forest, he controlled the transport system and mills necessary to process them, and he had amassed a fortune worth $80 billion today, making him the eighth-richest American of all time.

Frederick Weyerhaeuser, 1924.
MNHS collections

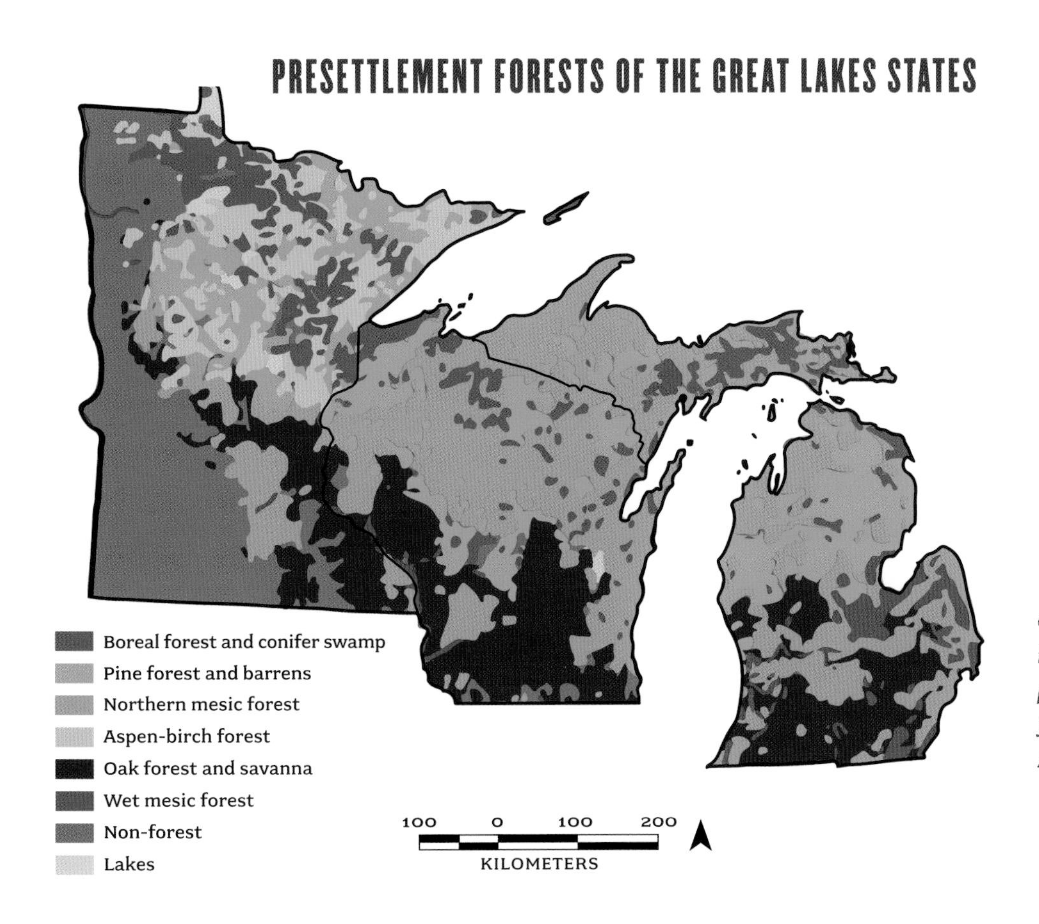

Great Lakes forests in the presettlement period. Graphic by Jennifer Osborne Anderson

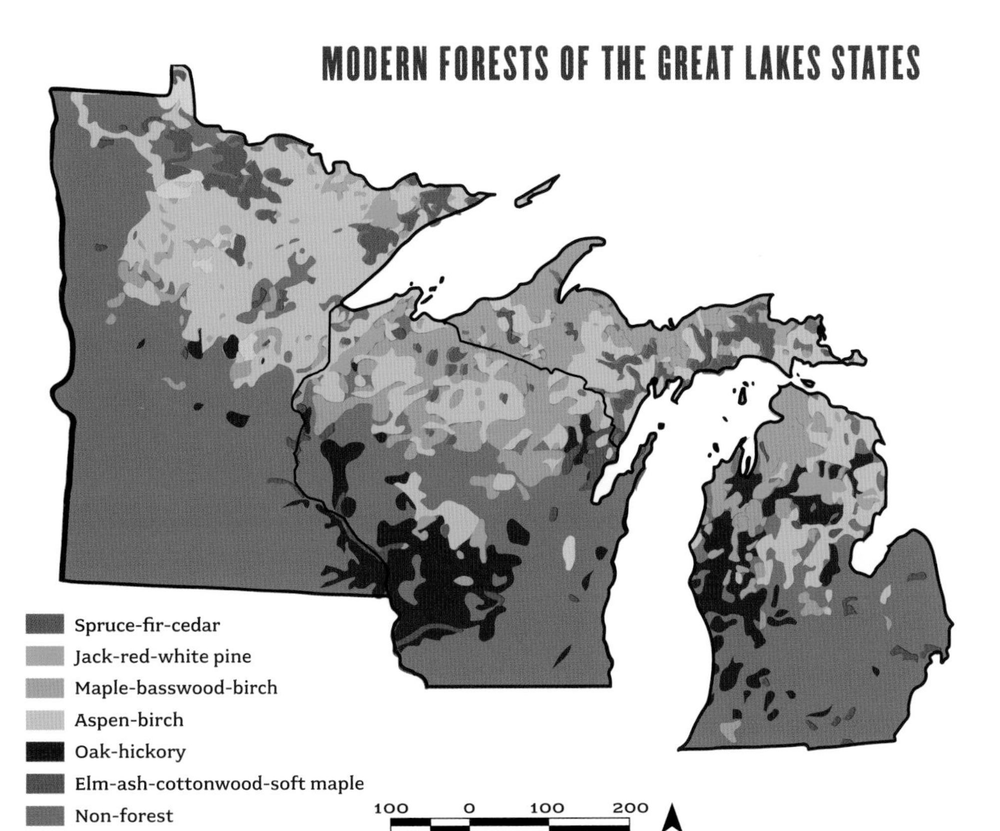

Great Lakes forests post-settlement. Graphic by Jennifer Osborne Anderson

TABLE 9–1: MINNESOTA CONIFEROUS FOREST HISTORIC CHANGES

	ACRES DOMINATED BY WHITE PINE FOREST	PERCENT REMAINING
1837	3.5 million	100
1936	224,000	6
1962	135,800	4
1990	67,500	3*

* Mostly second-growth forests (60–120 years old).

Source: White Pine Society, "DNR White Pine Management Plan—Critique." Wildlife Research Institute, http://www.whitepines.org/Critique.html.

the set-aside land. By 1902 thousands of acres of pine and arable lands had been transferred from the Ojibwe to the federal government and then sold to the public.

By 1920 the inexhaustible forest no longer seemed so endless. The yield was down to 576 million board feet that year, and lumbermen looked instead to the Pacific Northwest, with its enormous expanses of Douglas fir. In 1937, when the large sawmill in International Falls closed its doors, the era of big white pine logging was at an end. But during its reign, that industry had produced more than 67 billion board feet of lumber.

The boom made cities out of the settlements of Stillwater, Winona, Hastings, Red Wing, Minneapolis, St. Paul, and Duluth, spawned towns and farms across the state, helped build the railroads, and made Minnesota a prosperous industrial state. But the clear-cutting left a barren landscape in Minnesota's north country. Lumbermen wanted the trunk of the tree, not the tops or the branches. These sections were left as slash, dry scraps that were perfect kindling material. Euro-Americans frequently burned land to clear it; railroad sparks set the brush beside the tracks on fire. Thousands of land speculators, timber assessors, and sportsmen left smoldering campfires, and trash fires burned untended around logging camps. Not surprisingly, many fires swept through lumbering areas.

In 1894 a combination of extreme drought conditions, high winds, and small fires caused a catastrophic inferno at Hinckley and four other small towns nearby. Called the Great Hinckley Fire, the blaze incinerated the town, destroyed 480 acres, killed 418 people, and left only the roundhouse and the water tower in its path. Cutover land and dry seasons sparked numerous other fires in the region; one commentator noted 340 wildfires in 1908. Two other major fires occurred in autumn in the following years: the Spooner-Baudette Fire of 1910 and the Cloquet–Moose Lake Disaster of 1918.

A total of 453 people died in the Cloquet fire, and 52,000 were injured or displaced. Disastrous as they were, these fires helped awaken citizens, politicians, and businessmen alike to the need for fire protection and forest replenishment.

Logging continued at a reduced pace throughout the twentieth century and continues today. When cutting began, 3.5 million acres of Minnesota forests were dominated by white pine. By 1990, 98 percent of Minnesota white pine was gone, leaving 67,000 acres that included pine, most of it young growth. The decline has been greater here than elsewhere in the nation. For example, Minnesota originally had twice the white pine acreage of New Hampshire but now has less than a twentieth as much.

This Sebeka High School mural of loggers in the north woods was painted in 1938 by Richard Haines, a Minneapolis Art School graduate, through a program of the Works Progress Administration.

Recently the logging industry has adopted much more sustainable practices. Various state programs help provide training and certification to meet standards that include safe, productive, and environmentally responsible timber harvesting. White pine is no longer the primary tree cut as timber: oak and aspen are also desirable. In addition, the current law requires that when trees are cut down, they must be replaced with seedlings.

REGENERATION AND RESTORATION EFFORTS

Many Minnesotans watched the continual harvest of white pine during the peak years and were alarmed at the wreckage of the landscape. But many people saw the clearing of forests as a benefit to civilization. Others saw no need to conserve forests, since they did not believe timber could ever be in short supply. As one logger, quoted in Susan Flader's book *The Great Lakes Forest*, recalled when speaking about logging practices:

> When the spring come and the ice went out . . . the logging camp crew scattered. Some of the men went down with the drive. Others went home to families and started clearing land and planting and making 160 acres of stumps into a farm. . . . They left behind a sorry-looking land: miles of stumps and brush, piles of branches where the swampers and skidders trimmed the trees, roads that never growed up again. I suppose we should have burned our brush as we cut off the trees, so forest fires couldn't get going so easy. But we never knowed much about such things in them days. . . . The government shouldn't have let them companies take off the little trees that wouldn't make anything bigger than a two by four. Us loggers thought the big woods would last forever. I guess we can't expect the government to be much smarter than us.

Fortunately, a number of forward-looking citizens began to lobby for restoration and conservation. One of the most passionate, articulate, and persistent was General Christopher C. Andrews. Andrews had loved trees all his life, and during his time as US minister to Sweden, he learned about that country's forest conservation policy. From then on he lobbied Congress, testified before Minnesota legislators, and prepared detailed reports about managing timberlands and establishing fire protection. He was interested in scientific forestry and worked to establish a school of professional forestry. He wrote and spoke repeatedly about preserving sections of the forest and replanting after trees were harvested. His message often fell on deaf ears, yet he persisted through his long life.

Minnesota's two national forests, the Chippewa and the Superior, began as Andrews's ideas and became realities as he garnered support. The Chippewa National Forest in north-central Minnesota was set aside in 1902, the Superior National Forest in 1909. Andrews lobbied for state land as well and was able to have Burntside Forest set aside in 1904. He worked with the Minnesota State Forestry Association to draft a forest reserve bill. As a result of his dramatic lobbying and testifying, in 1911 Minnesota passed a forestry law that concentrated all forestry and firefighting tasks into the office of the state forester. Before Andrews died at age eighty-two in 1922, he was able to see a number of improvements in forest policy, many the result of his continued pleas. In 1943 the state named its newest forest in his honor: the General C. C. Andrews State Forest in Pine County.

In 1909 President Theodore Roosevelt signed a proclamation officially creating the Superior National Forest, located in the Arrowhead region of the state. Much of the original 644,114 acres was cutover and burned-over lands—a border area that had become "lands that nobody wanted." Today, at 3.9 million acres, the forest has almost quadrupled in size. During the Great Depression, members of the Civilian Conservation Corps carved out miles of trails there and collected thousands of cones. They inventoried the forests and planted pine plantations, still visible today. The Wilderness Act of 1964 set aside the area now designated as the Boundary Waters Canoe Area (BWCA) as a wilderness reserve of 1.09 million acres within the Superior National Forest.

Individuals, conservation groups, and government agencies have spearheaded a variety of replanting schemes throughout the state

WHAT MINNESOTANS HAD TO SAY ABOUT WHITE PINE

The wonderful wood of the white pine drew the lumber industry to Minnesota. They clearcut thousands of acres of this majestic tree, leaving the landscape forever changed. The few that remain should be treasures that should be protected.

D. E.

over the years. Although all these have been of benefit to the white pine inventory, most have been hampered by a number of factors. White pine blister rust was introduced on seedlings brought from Europe in the early 1900s and remains a problem. The large deer and rabbit populations in Minnesota graze on young white pines, killing many. In addition, white pine seeds require disturbed soil to germinate. Forest fires and tilling are two ways of disturbing the soil, but forest fires have been suppressed for decades, and tilling is not practiced in the forests. Also, much of the land that was once forested either is developed or has become farmland, limiting the areas available for trees to establish themselves.

In addition, the slow-growing trees need sunlight to thrive. Hardwood trees and undergrowth block the sun's rays, so white pine has trouble producing new plants in densely vegetated areas. Finally, reforestation is an expensive task. According to the University of Tennessee Agricultural Extension Service's document "Tree Crops for Marginal Farmland: White Pine, with Financial Analysis," the cost of planting and maintaining one acre of white pine forest in the year 2000 was approximately $240.

Today the state of Minnesota, local government, industry leaders, and individuals have a number of programs in place that are making good progress in restoring the native ecosystem of white and red pines. Each year the Minnesota Department of Natural Resources (DNR) plants 1 million seedlings on state lands. In addition, the DNR has formed a technical team to develop a plan that will aid white pine growth. Its scientists are studying ways to improve rust-resistant seedlings.

Through the Environmental Quality Incentive Programs, the USDA's Natural Resources Conservation Service (NRCS) provides landowners and other agriculture producers with financial and technical assistance to address natural resources concerns. Under

Gunflint Pines *by Anna C. Johnson*. Courtesy Johnson Heritage Post Art Gallery, Grand Marais, MN. Used with permission.

THE LOST FORTY

THOUGH MINNESOTA has very little virgin forest left, a choice bit of acreage remains in the Chippewa National Forest. This easily accessible land, the "lost forty," is home to one of the last old-growth sites of red and white pine in Minnesota. Some of the trees are 350 years old and between twenty-two and forty-eight inches in diameter. Here live numerous birds, including woodpeckers, bald eagles, and hawks—ninety species have been recorded in the preserve. Many small and large mammals, black bears, shrews, red squirrels, and weasels also live in these woods. A one-mile, self-guided trail winds its way through the tall pines.

Our gain is due to an oversight by early surveyors. In 1882 a federal crew led by Josiah A. King was finishing its work in the north woods. The chilly weather, the November winds, or the desolate swamp around them confused the men. They marked the 144 acres incorrectly, listing it as Coddington Lake, not the virgin pine wood that it was. Because it was mistakenly marked as water, the land went uncut by the logging companies. The acreage has since been declared a Scientific and Natural Area by the Department of Natural Resources and is managed to maintain its old-growth character. The "lost forty" bears witness to our heritage; these pines were seedlings when the pilgrims came to America. Take a virtual tour at the DNR site: http://www.dnr.state.mn.us/snas/detail.html?id=sna01063.

The "lost forty" in the Chippewa National Forest. John Gregor/ColdSnap Photography

these arrangements, individuals who own forest lands meet with experts from the NRCS to develop a plan for conservation and reforestation for white and red pine.

The Nature Conservancy has partnered with state, national, and county institutions in the Sand Lake/Seven Beavers area of northeastern Minnesota to restore and manage the forest. Here these groups are working as partners with a mixture of private owners to develop a collaborative approach to forest management and conservation. This project is expected to become a model for keeping forests productive and healthy for people.

Near Bemidji, the Red Lake Band of Ojibwe is replanting 50,000 acres of tribal land that were misused by the federal government starting in the early 1900s. The band's goal is to reforest about 1,000 acres annually for the next fifty years, using the types of trees that once grew on the land. It has set up a Forest Development Center that is open to students for historical, cultural, and natural resources education. The center contains three state-of-the-art greenhouses, a technologically advanced nursery and seed bank, and a laboratory and testing facility.

Many smaller efforts, such as the Minnesota Twins' Break a Bat, Plant a Tree partnership with the DNR, provide a significant number of new plantings each year. Now in its eighth year, the partnership results in the planting of a hundred trees in a Minnesota park or recreation area or along a state trail every time a Twins pitcher breaks the bat of someone on the opposing team (the trees are planted in the spring that follows the baseball season). During the 2010 season, the pitchers broke 180 bats, resulting in 18,000 new trees planted, many of them red and white pine.

WHAT MINNESOTANS HAD TO SAY ABOUT WHITE PINE

Lumbering of white pine decimated virgin forests and earned fortunes for lumber barons and railroads. But it also paved the way for numerous ethnic and cultural groups to immigrate to northern Minnesota.

DEBORAH S.

WHITE PINES *and the* ENVIRONMENT

WHITE PINES fill important ecological roles. They grow across a range of forest conditions and are especially suited to the northeastern part of the state. Many animals rely on the tree for food and shelter. Rodents, bears, and birds eat the pine seeds. Eagles and ospreys nest in the tall white pines. White pines hold 77 percent of Minnesota's osprey nests and 80 percent of the state's bald eagle nests. The rough bark is easy for black bears to climb, making white pine a favorite for mothers and cubs. Several species of butterflies and moths feed on Eastern white pine.

Restoring and retaining the white pines in the state will improve the quality of water flowing through any watersheds in their location, so that our rivers and lakes will be more pristine. The dense branches of conifers do a better job of shading streams in summer months and slowing snow melt in spring, thereby helping to prevent the erosion that fouls waterways. Restoration will maintain and improve the soil's ability to support and retain water. In addition, growing white pine will increase carbon storage because long-lived trees can hold carbon for generations without releasing it in decomposition.

White pine along the Gunflint Trail. Mary H. Meyer

DID YOU KNOW?

- The **SCIENTIFIC NAME** for white pine is *Pinus strobus*, translated literally from the Greek as "pine cone," for its long, narrow cone.
- **COLLOQUIAL NAMES** for white pine are old field pine, mast pine, cork pine, and soft pine.
- White pine came to Minnesota about **7,000 YEARS AGO**, following the last Ice Age.
- White pines were once so plentiful that a person could travel from the **ATLANTIC COAST TO MINNESOTA** and seldom be out of sight of them.
- The tree's **BLUE-GREEN NEEDLES** grow in bunches of five, making them easy to identify.
- **MINNESOTA** has a Pine County, a Pine Lake, a Big Pine Lake, a Pine City, a Pine Point, a Pine River (city), a White Pine Township, and a Pine Lake Township.
- The white pine is a relatively **LONG-LIVED TREE**, with an average maximum age of probably more than 500 years.
- **BEFORE LOGGING BEGAN**, white pine was found in every Minnesota county east of the Mississippi River from Minneapolis north to Canada.
- Minnesota's largest white pine, found in Itasca State Park, is **131 FEET TALL AND 180 INCHES** in circumference.
- White pines are the **LARGEST CONIFER** in Minnesota.
- Pine resin (sap) was used by native peoples to **WATERPROOF** baskets, boats, and pails.

Cones of white pines.
David L. Hansen, University of Minnesota

10
WILD RICE

**Wild rice, or manoomin,
is a sacred food and medicine integral
to the religion, culture, livelihood,
and identity of the Anishinaabeg [Ojibwe]. . . .
Whatever happens to the land and to manoomin
happens to the Anishinaabeg.**

ERMA VIZENOR, TRIBAL CHAIRWOMAN, WHITE EARTH NATION, 2008

> [Wild Rice] is the most valuable of all the spontaneous productions of that country [Great Lakes area]. Exclusive of its utility as a supply of food for those of the human species who inhabit this part of the continent and obtain it without any trouble than that of gathering it in, the sweetness and nutritious quality of it attracts an infinite number of wild fowl of every kind which flock from distant climes to enjoy this rare repast.
>
> JONATHAN CARVER, ENGLISH EXPLORER, 1766

Wild rice (*Zizania palustris*), or manoomin, as the Ojibwe people call it, is a grain native to North America. Not directly related to Asian rice (*Oryza sativa*) but a distant cousin, wild rice is a persistent annual aquatic grass that grows in the cool waters of northern Minnesota. Found primarily in the Great Lakes region, wild rice has been used for human consumption for more than 2,000 years.

Some studies indicate that it was used for food about 5,000 years ago. Wild rice was utilized as a major part of Native American subsistence throughout the Woodland period (2500–0 BCE) and has been found in layers of the earth dated back 12,000 years.

Minnesota has more acres of natural wild rice than any other state in the country, and the plant has been historically documented in fifty-five of Minnesota's eighty-seven counties. Cass, Aitkin, Itasca, and St. Louis Counties contain the most acreage. In 2008 a state inventory found wild rice stands in approximately 1,286 lakes, rivers, and streams.

At 7,400 acres, Nett Lake, on the Bois Forte Indian Reservation in St. Louis County, is the largest and most abundant wild rice lake.

Manoomin *(wild rice), on Perch Lake in Sawyer, Minnesota, on the Fond du Lac Ojibwe Reservation.* Photo by Ivy Vainio

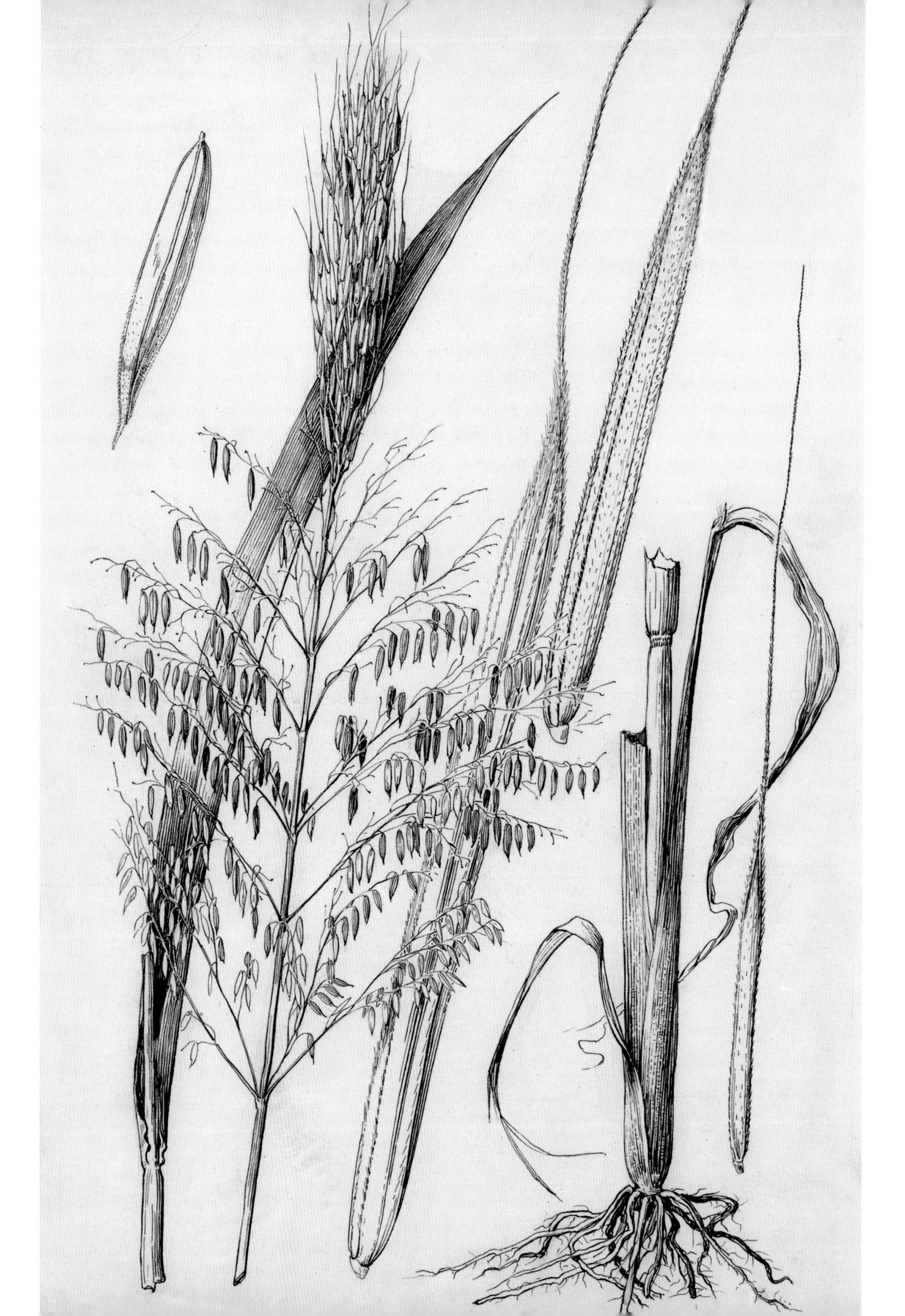

Its location near the top of three major North American watersheds, away from urban areas, isolates it naturally from sources of pollution. The Bois Forte Band has never allowed the use of fertilizers, pesticides, or outboard motors on the lake, keeping the water clear and wild rice plentiful.

Today wild rice is Minnesota's official state grain, and numerous of the state's towns and bodies of water have names that include the word "rice" or "manoomin." There are 123 Rice Lakes, a Rice Lake Preserve, Rice Creek, three Rice Rivers, the town and county of Mahnomen, and Wild Rice Lutheran Church of Twin Valley. In fact, it is thought that no other plant has contributed to more geographic names in all of North America.

Why is the plant labeled wild "rice" when it is a grass? The early European explorers saw it growing in water and observed the local people harvesting it. Because the habitat and harvest reminded them of rice paddies, they assumed it was a "wild" rice.

The plant was a chief food for the many tribes in the Great Lakes region (including the Menominee, Ottawa, Potawatomi, Meskwaki, Ho-Chunk, Ojibwe, and Dakota). It was especially valuable because it could be stored indefinitely after being harvested and dried. These tribes all had a diversified economy and relied on wild rice for food and trade. It was so essential to the Dakota of northeastern Minnesota and the Great Lakes Ojibwe bands that the two battled for more than a century over access to the rich wild rice territories of northern Wisconsin and Minnesota.

Early European explorers assigned much importance to wild rice as well. Their journals contain many references to the plant they found growing in the lakes and rivers they encountered. Because it was difficult to ship the food they usually ate, they came to rely on the local food supply. As a staple food of the voyagers, wild rice helped the region's fur trade flourish, providing nourishment to

Wild rice botanical drawing from the Hitchcock-Chase Collection of Grass Drawings. Courtesy the Hunt Institute for Botanical Documentation, Carnegie Mellon University

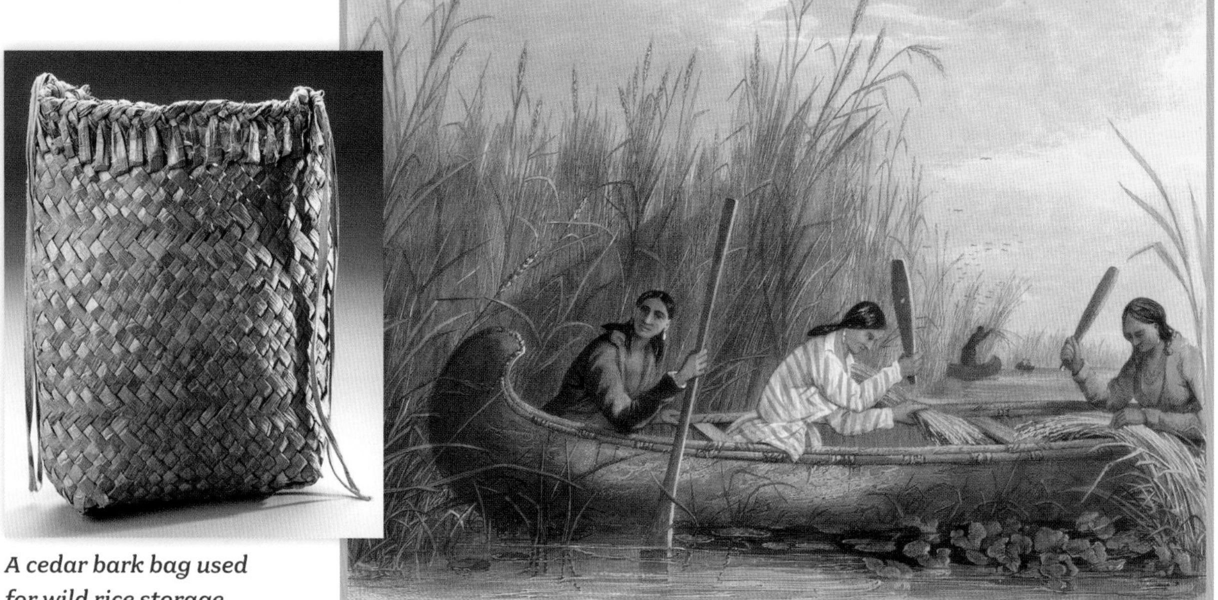

A cedar bark bag used for wild rice storage. MNHS collections

Gathering Wild Rice, 1853, by Captain Seth Eastman. Newberry Library, Chicago, Illinois, USA/Bridgeman Images.

traders and hunters. Because careful preparation and storage made it virtually imperishable when kept dry, it was invaluable during the long winters. So important was this plant to Euro-Americans in the region that in 1837 one resident of Fort Frances, just north of International Falls, noted, as quoted by Thomas Vennum in *Wild Rice and the Ojibway People*: "Our sole dependence and principal food for the winter: Wild Rice may be truly called the staff of Life, of this post." Many immigrants to Minnesota adopted the grain as a basic supplement to their foodstuffs, and some learned to hand harvest it.

The Ojibwe became the suppliers of wild rice to the explorers and immigrants. They received other goods in trade, and in later years received cash payments.

WILD RICE LIFE-SPAN

After ice is out in the spring, wild rice, an annual plant, begins to grow from seeds left in the lake during previous years. The plants grow best in water between one and two feet deep, but they can grow in shallower water or in water up to three feet deep. In late May and early June, wild rice is in the "submerged leaf" state, when a cluster of up to four underwater basal leaves form. (Basal leaves grow on the

WILD RICE AS A SPIRITUAL FOOD

WILD RICE has been a central feature in the lives of the Dakota and Menominee (who took their name from the plant), as well as the Ojibwe. For the Ojibwe, or Anishinaabe, wild rice is much more than a source of nutrition and has a spiritual and cultural significance. It is featured in ceremonies and legends, and its use is regulated by taboos and proscriptions. The size of the harvest, damage to the crop, and the pace of its growth—all are considered to have a supernatural essence.

According to the sacred story, the creator, Gichi-Manidoo, guided the Anishinaabe on their westward journey to "the food that grows on water." The creator's gift was nutritious in itself but also was home to many birds and waterfowl that provided sustenance. This food is still remembered in the wild rice harvest, in ceremonies, and in thanksgiving feasts. The Anishinaabe named it manoomin, which means "good berry."

Knocking sticks are used to bring the rice into the boat. Great Lakes Indian Fish & Wildlife Commission

continued next page

Jim Northrup and his son Jimmy Northrup manoominike *(harvesting wild rice) on Perch Lake, 2013, in Sawyer, Minnesota, on the Fond du Lac Ojibwe Reservation.* Photo by Ivy Vainio

WILD RICE AS A SPIRITUAL FOOD *(continued)*

The seasons of rice became a way of marking time. Events were described as occurring before or after wild ricing. For example, American Indians generally named their months according to some aspect of their economy. Thus, June is the strawberry moon; July, the raspberry moon. Invariably, August or September was designated the month of collecting wild rice. In the Ojibwe language, the phrase "manoominike-giizis" signals the harvest season, a time for celebration as well as a time of labor. In this season, families, separated by place and time, come together again.

The Anishinaabe believe that wild rice will always grow where they live. The plant is present during periods of mourning and at initiations, feasts, and celebrations. One elder noted that wild rice and water were the only two items required at every ceremony.

lowest part of the plant.) Wild rice surfaces by mid-June and initially floats on top of the water, with ribbon-like leaves that form a vast, leafy mat. During this period, known as the "floating-leaf" stage of wild rice growth, changes in water level, the amount of cloudiness, or extreme water disruption or flow can uproot the plant. Wild rice plants also suffer when competing with other vegetation in the lake, including water lilies, water shield, and pickerel weed. Sometimes these other plants begin growing in the spring before wild rice and take over the water surface. By the end of June, the aerial shoots have begun to develop. They grow into August, reaching a height of two to eight feet above the water surface. When the water is shallow and plant density is low, multiple shoots, ten or more, can develop.

In late July or early August, flowering begins. Both male and female flowers grow on the same stalk, the female flower above the male. The female flowers are the first to emerge from the stalk, with the male flowers following three or four days later. This staggered opening encourages cross-pollination. The plants are wind pollinated, with seeds on a single stalk reaching maturity over a ten- to fourteen-day period. Ripening is affected by water depth, weather conditions, sediment type, and other factors. Even with the same bed of plants, the seeds ripen at various times; thus, harvesting is done repeatedly in the same area, generally from late August into September.

The seed that is not harvested will fall into the water and sink into the mud, remaining dormant until warming water and low oxygen conditions stimulate germination. This extended dormancy allows wild rice to survive an occasional crop failure. Some seeds will germinate the next year, but they can remain viable for five years.

Wild rice plants. David L. Hansen, University of Minnesota

TRADITIONAL HARVESTING

Traditional wild rice harvesting consists of five steps: knocking, drying, parching, hulling, and finally winnowing. The harvest is done with two people working in a canoe. One guides the canoe while the other knocks the stalks to harvest the rice. Two knockers (sticks of specified size and weight) are used, one to pull the stalks toward the canoe and the other to lightly brush the stalks so that the grains fall into the canoe. If the grain is not ripe, the wild rice will generally not fall when it is knocked, requiring harvesters to repeat the process over a period of several weeks.

On shore, the grains are spread out in a shady spot in a single layer, generally on canvas, sheets, or birch bark. The stalks and leaves are removed and the grains stirred every few hours to remove the debris and ensure that they dry evenly. Once dry, the wild rice grain is roasted—or parched—in an iron kettle over a small wood fire, for about thirty minutes to an hour while being constantly stirred by paddles to avoid burning. During the roasting the kernels are

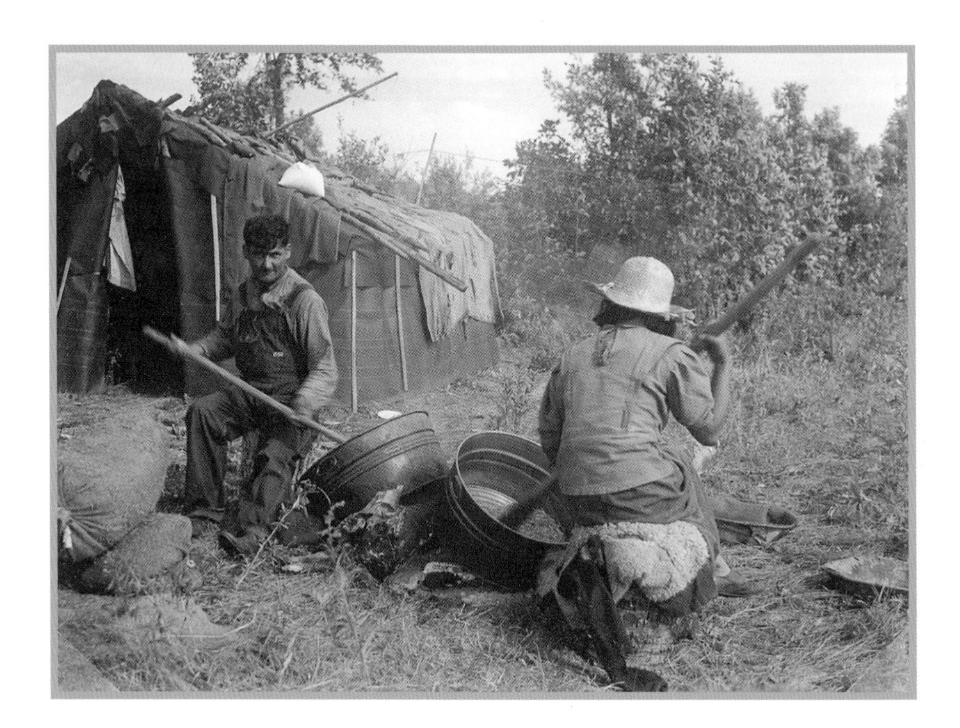

Paul Buffalo and his wife parching rice, 1937. Bureau of Indian Affairs, Minnesota Area, courtesy National Archives

snapped from time to time to determine if the inside has become glossy. A shiny appearance indicates that the rice has dried enough to have the hulls removed.

Once the kernels are roasted, the inedible hulls and chaff are removed using one of two methods. Usually this task is done in a small pit in the ground lined with wood, called a bootaagan. There the wild rice grain is treaded ("jigged"), or danced on. The jigger holds onto poles placed alongside the pot; these poles help him keep his balance and also regulate how heavily he is treading the rice. Wearing special, new moccasins, the dancers tread on the rice in a circular motion.

The other method is to pound the rice, using heavy wooden pestles that are repeatedly dropped onto the rice. Generally five or six feet in length, these pestles put enough weight onto the rice to remove the hulls without requiring any extra force.

The last step of the harvesting process, winnowing, removes the chaff from the parched kernel. The grains are put in a tray, usually made of birch bark, and tossed slightly so that the chaff is blown away naturally by the wind. The tossing is repeated several times to remove all of the chaff. After this final step, the wild rice can be stored in containers for later use.

Winnowing the parched and hulled rice, 1935. Bureau of Indian Affairs, Minnesota Area, courtesy National Archives

Different aspects of the harvesting procedure were handled by different genders and ages; today many of these duties can be shared between men and women. Both men and women knocked rice into the canoes. Boys and young men typically handled the arduous task of thrashing or jigging. Skilled older women took care of the winnowing.

In the past, as ricing season neared, Ojibwe families moved from their summer homes and set up camps along wild rice lakes and rivers. All looked forward to these times, which involved social components as well as work. The families remained in camp for the months of ricing and processing. In the 1940s, as Ojibwe acquired cars, day trips to the wild rice lakes were convenient, and the camping tradition gradually faded. It has been kept alive in memory and by many families. Historically the camps helped strengthen community ties and communication between the generations. To teach the younger generation about the importance of hand harvesting and the wild rice culture, several reservations continue to hold wild rice camps to demonstrate proper techniques.

COMMERCIAL PRODUCTION

For more than a hundred years, individuals and groups have seeded or reseeded lakes to improve wild rice stands. Indeed, many Ojibwe collected seed and established new stands in suitable lakes. In the 1940s the Minnesota Department of Natural Resources, then called the Department of Conservation, seeded lakes and streams in efforts to improve wild rice stands in the state.

As early as 1853, Oliver H. Kelley, a self-taught expert in Minnesota farming and cofounder of the National Grange, was the first to propose cultivating wild rice as a crop, but nothing came of his suggestion. Not until the 1950s did anyone in Minnesota begin cultivating and cropping wild rice. Near Crosslake in Crow Wing County, two farmers, Jim and Gerard Godward, flooded a one-acre field and planted it with seed collected from a lake—thus the first cultivated wild rice. Their success encouraged them to flood and plant larger areas. For years they and their families experimented with harvesting processes and planting methods, all the while increasing yields. Others soon followed, and the industry expanded. When Uncle Ben's Rice, owned by the Mars Company, developed a wild and white rice blend in the 1960s, Minnesota's cultivated rice industry grew rapidly.

Currently wild rice is cultivated in the northern third of Minnesota. Aitkin, Beltrami, Clearwater, and Polk Counties usually produce a majority of the state's crop. Cass, Crow Wing, Itasca, Koochiching, Lake of the Woods, Pennington, and Red Lake Counties also contribute to the state's production. Through the years, farmers have improved equipment and methods, now harvesting with combines over the soggy peat bed.

Most paddies (rice fields) are created on nonarable land that has a high peat content. Farmers make permanent dikes that will give consistent water depth in their paddies so the wild rice plants can grow in an aquatic environment. In March or early April, the paddies are flooded with six to eight inches of water, and the wild rice seed is planted. To ensure that the rows are not overcrowded, once the plants reach maturity they are thinned using a propeller-driven air-

Cultivated (left) and wild harvested wild rice.
Mary H. Meyer

WHAT MINNESOTANS HAD TO SAY ABOUT WILD RICE

It is the only cereal grain native to North America; its domestication history is centered in Minnesota (and the U of M). Its production has evolved into a multimillion-dollar industry that directly benefits some of the most economically depressed areas of the state.

PETER I.

boat equipped with cutters. Three weeks before harvest, the fields are drained to allow them to dry out before specialized combines harvest the crop. Due to the development of varieties that do not shatter their seed as easily as the lake types, growers need to harvest the fields only once.

In addition to the growing processes, farmers must deal with other issues: challenges in maintaining correct and consistent water depths, the problems of shattering seed during strong winds, seed losses due to blackbirds, and fluctuating temperatures that affect the development of the grain. Diseases, including brown spot and stem rot, are also common problems, as is insect damage, especially from the wild rice worm feeding on the developing grain.

In the 1950s the University of Minnesota began studying wild rice production on a small scale. However, it was not until 1973 that the state of Minnesota, at the urging of the fledgling wild rice industry, appropriated funds for a comprehensive research program. The challenges the growers and the university faced in making a commercial crop viable were numerous. For one, wild rice grown in a natural stand matures at widely different times, so several harvests must be made. In addition, the seed shatters when ripe, tossing some of the crop into the water. However, with the new nonshattering varieties, seed loss is not as much of a problem as it was previously. In nature, this shattering helps ensure ongoing crops, but it is not useful to commercial growers. Also, the seed must be stored in conditions similar to a lake bottom to remain viable. Most of the university's work has taken place at the North Central Research and Outreach Center in Grand Rapids, but much was also done at the St. Paul campus and in the growers' fields.

Despite the difficulties wild rice presents, the University of Minnesota, in cooperation with growers, has successfully introduced many new varieties. The earliest one, Johnson, was released in 1968 as the first nonshattering variety. Voyager, with a short to medium height and early maturity, was added in 1983. A high-yielding variety, Purple Petrowski, is resistant to shattering and fungal disease and was released in 2000. Barron was released in 2014.

With between 13,000 and 16,000 acres planted to commercial wild rice yearly, Minnesota has become one of the leading producers of cultivated wild rice. Initially California farmers bought seed from Minnesota. Now those in California have a production about equal to that of the North Star State. In 2015 Minnesota farmers produced 7 million finished pounds of cultivated wild rice; the annual yield of wild rice from Minnesota rivers and lakes fluctuates between 200,000 and 500,000 pounds of finished wild rice. Most of the state's cultivated wild rice is used in wild rice/white rice blends.

Commercial paddy-grown or cultivated wild rice and handpicked or hand-harvested wild rice are both available for sale in Minnesota. In many states, only cultivated wild rice is available. Readily available hand-harvested wild rice is unique to Minnesota and other states where it grows naturally and is usually more expensive due to the labor involved in collecting and processing the grain.

Processing methods can affect the color, firmness, and final quality of wild rice. Visually, cultivated rice is usually more uniform, with a darker, almost black color and a firmer grain. Hand-harvested grains of wild rice are lighter brown, less uniform, and softer. These variations result from the different processing methods and the variability of the plants from which the grain is harvested. Cooking time is shorter for hand-harvested wild rice. Some people prefer the texture and taste of hand-harvested rice, while others cannot distinguish between the two growing and processing methods when eating the final cooked product.

TABLE 10—1: WILD RICE VARIETIES

RELEASED BY THE MINNESOTA AGRICULTURAL EXPERIMENT STATION
Johnson, 1968; M1, 1970; K2, 1972; M3, 1974; Netum, 1978; Voyager, 1983; Meter, 1985; Franklin, 1992; Purple Petrowski, 2000; Barron, 2014
RELEASED BY THE MINNESOTA CULTIVATED WILD RICE COUNCIL
Itasca, 2002; Itasca Cycle-12, 2007

Early exploration of wild rice had a broad scope. University researchers studied agronomics, seed storage, plant development, breeding, soil fertility, insect and disease control, harvesting methods, processing, marketing, and nutrition. The present focus of the university's research, much of which is supported by grants from the Minnesota Cultivated Wild Rice Council, is three-pronged. Researchers are working to develop varieties that resist shattering and disease, to increase yields and stem sturdiness, and to reduce seed dormancy. Additionally, research is directed toward finding seeds that withstand longer storage times. Finally, researchers are focusing on soil fertility and best management practices for the application of nitrogen.

WILD RICE AND THE LAW

Since the 1800s, wild rice has been regulated and defined by a tangle of laws and regulations, federal, state, and tribal. As Europeans moved into Minnesota territory, they expected to acquire land. On July 29, 1837, in St. Peter, the Ojibwe and Dakota people ceded land to the United States in central Minnesota and northern Wisconsin, but they retained rights to hunting, fishing, and gathering in those areas. Wild rice was their main foodstuff; thus, they negotiated to always have access to wild rice lakes.

Other treaties followed in the 1850s and 1860s, forcing the original inhabitants to give up their ancestral lands. For government officials, and European Americans in general, the Dakota and Ojibwe were impediments to the growth of the territory. When the Indians were removed from their lands, the territory would be developed and all but the Indians would profit. Land speculators and immigrants were flooding to the area, seeking not just land but also access to the lakes and streams.

Different bands, negotiating as sovereign Indian nations, sought to stay in the heart of their lands; the Ojibwe, who lived in forested areas less suited to agriculture, were more successful. Seven reservations were established for the Ojibwe in the northern section of the state and one was established for the Dakota in the south. Many bands found that their subsistence needs were not being met. Often, their access to game and to lakes outside of the reservation—guaranteed in the treaties they had signed—was dependent on the goodwill of the newcomers. After the Dakota War of 1862—which started largely because Dakota people were starving—most of the Dakota were sent to reservations in South Dakota and Nebraska. Over the years, though, they returned, and in the 1930s four Dakota communities were established in southern Minnesota. By the middle of the twentieth century, the newcomers had begun building cabins on the lakes, clearing the wild rice from the shoreline, and using the lakes for recreation. In the 1930s commercial ricing interests began using machines that harvested by clear-cutting, thus devastating rice stands. The Ojibwe were understandably frustrated with the premature harvesting they observed taking place at non-Indian hands.

The state of Minnesota and the federal government have frequently been at odds in handling wild rice issues. For example, in 1926 Congress enacted Public Law No. 418, which created a protected area, the Wild Rice Lake Indian Reserve in Clearwater County, on the White Earth Reservation, intended for the "exclusive use and benefit of the Chippewa Indians of Minnesota." However, the state of Minnesota wanted this space to be a game refuge and hunting ground. Therefore, Minnesota began condemnation hearings in 1934 to acquire the property. At that point, a legal battle began between Minnesota and the federal government. In the spirit of the New Deal, Congress reaffirmed its protection of the Wild Rice Lake Indian Reserve in 1935 but used money from the Ojibwe trust funds in the federal treasury to implement the reserve.

In 1939 the Minnesota legislature passed Statute 84.09, Chapter 231, Conservation of Wild Rice, which acknowledged native treaty rights, giving the "Indians exclusive rights" to harvest the wild rice crop upon all public waters within the original boundaries of the White Earth, Leech Lake, Nett Lake, Vermillion, Grand Portage, Fond du Lac, and Mille Lacs Reservations. In addition, machine harvesting

WHAT MINNESOTANS HAD TO SAY ABOUT WILD RICE

"Wild rice was a staple food source for Minnesotans (especially in the north) for hundreds of years, and it remains central to Ojibwe culture" (Thomas Vennum, *Wild Rice and the Ojibway People*, 1988). More "wild" wild rice grows in Minnesota than in any other state. There is evidence that wild rice harvesting led to a population increase in the late Woodland period (T. Lofstrom, "The Rise of Wild Rice Exploitation and Its Implications for Population Size and Social Organization in Minnesota Woodland Period Cultures," 1987).

ETHAN PERRY, AGRICULTURAL UTILIZATION RESEARCH INSTITUTE

WHAT MINNESOTANS HAD TO SAY ABOUT WILD RICE

It was a staple for Native Americans, enabling them to settle and survive in Minnesota. Later it was shared with Europeans and exists today as a special Minnesota food.

JULIE D.

was banned on public waters, and a license was required to rice. The commissioner of conservation was to be responsible for developing and carrying out this program. The Ojibwe objected to portions of this plan because the commissioner was given the power to say when ricing could begin, rather than traditional Ojibwe rice camp leaders, and because purchasing a license at times required them to travel long distances.

In 1988 several Ojibwe filed suit in *Wabizii v. Busch Agricultural Resources*, claiming that Busch Agricultural Resources engaged in false and misleading advertising with its Onamia Wild Rice, a paddy rice whose package depicted two Indians in a canoe knocking wild rice. The case was settled out of court, but a short time later Minnesota passed Statute 30.49, mandating that Minnesota paddy wild rice farmers label their products as "paddy rice."

Reasserting their rights to the ceded territory, in 1990 the Mille Lacs Band of Ojibwe filed a lawsuit against the state of Minnesota. Two years later the Fond du Lac Band filed a separate lawsuit, seeking the same recognition. The Mille Lacs case moved through the district courts and reached the US Supreme Court in 1999. The Court decided in the Mille Lacs Band's favor, once again stating their rights to fishing, hunting, and gathering as enumerated in the 1837 treaty.

In 1984 the Great Lakes Indian Fish and Wildlife Commission was founded. This intertribal organization of eleven Ojibwe bands in Minnesota, Wisconsin, and Michigan governs the territory ceded in 1837, 1842, and 1854, managing off-reservation treaty seasons and protecting treaty rights and natural resources.

In 1973 Minnesota passed a law limiting sulfate discharges into wild rice–producing waters during periods when the rice might be susceptible to damage: the limit was no more than ten milligrams of sulfate per liter of water. Sulfate, a form of sulfur that occurs naturally, is also a by-product of industrial activities, such as mining and wastewater treatment. The law was based on research done in the 1930s and 1940s by state biologist John Moyle, who discovered that no large rice stands grew in waters that were high in sulfate. Researchers, mining companies, and some legislators have questioned the science behind the law. In 2015 the Minnesota Pollution Control Agency proposed a new standard: rather than relying on one sulfate level for all rice lakes, sulfate levels should be calculated for each body of water in which wild rice grows.

Ricing on public waters is regulated by the state's Department of Natural Resources (DNR), which sets the dates, times, and license fees and specifies the equipment that can be used. Native wild rice within the boundaries of the Bois Forte, Fond du Lac, Grand Portage, Leech Lake, and White Earth Reservations is managed by the respective reservation wild rice committees. These committees set the opening dates, times, and rules of harvest.

WILD RICE AND NUTRITION

Wild rice is an excellent grain, with a chewy texture and a nutty flavor; it is low in fat, salt, and cholesterol but high in fiber and protein and is gluten-free. One cup of cooked rice provides about 6.5 grams of protein but contains only 166 calories. Wild rice also provides folic acid, niacin, potassium, zinc, and several B vitamins. Recent research at the University of Minnesota indicates that wild rice has antioxidant properties and contains many phytochemicals. Animal studies in China have found that wild rice was effective in lowering cholesterol levels.

Cooked wild rice. iStock.com/AntiGerasim

TABLE 10-2: NUTRITIONAL COMPARISON OF WILD RICE AND FIVE OTHER CEREALS OR GRAINS

		WILD RICE, RAW		RICE, WHITE, MEDIUM-GRAIN, RAW, ENRICHED		RICE, BROWN, MEDIUM-GRAIN, RAW	
NUTRIENT	UNIT	VALUE/ 100 G	1 CUP = 160.0G	VALUE PER 100 G	1 CUP = 195.0G	VALUE PER 100 G	1 CUP = 190.0G
Energy	kcal	357	571	360	702	362	688
Protein	g	14.73	23.57	6.61	12.89	7.5	14.25
Total lipid (fat)	g	1.08	1.73	0.58	1.13	2.68	5.09
Carbohydrate, by difference	g	74.9	119.84	79.34	154.71	76.17	
Fiber, total dietary	g	6.2	9.9	1.4	2.7	3.4	6.5
Calcium, Ca	mg	21	34	9	18	33	63
Iron, Fe	mg	1.96	3.14	4.36	8.5	1.8	3.42
Magnesium, Mg	mg	177	283	35	68	143	272
Phosphorus, P	mg	433	693	108	211	264	502
Potassium, K	mg	427	683	86	168	268	509
Sodium, Na	mg	7	11	1	2	4	8
Zinc, Zn	mg	5.96	9.54	1.16	2.26	2.02	3.84
Vitamin C, total ascorbic acid	mg	0	0	0	0	0	0
Thiamin	mg	0.115	0.184	0.578	1.127	0.413	0.785
Riboflavin	mg	0.262	0.419	0.048	0.094	0.043	0.082
Niacin	mg	6.733	10.773	5.093	9.931	4.308	8.185
Vitamin B-6	mg	0.391	0.626	0.145	0.283	0.509	0.967
Folate, DFE	µg	95	152	386	753	20	38
Vitamin B-12	µg	0	0	0	0	0	0
Vitamin A, RAE	µg	1	2	0	0	0	0
Vitamin A, IU	IU	19	30	0	0	0	0
Vitamin E (alpha-tocopherol)	mg	0.82	1.31	0.11	0.21	0	0
Vitamin D (D2 + D3)	µg	0	0	0	0	0	0
Vitamin D	IU	0	0	0	0	0	0
Vitamin K (phylloquinone)	µg	1.9	3	0	0	0.536	1.018

NUTRIENT	UNIT	WHEAT FLOUR, WHOLE-GRAIN DEBRANNED VALUE/ 100 G	1 CUP = 120.0G	OAT FLOUR, PARTIALLY YELLOW VALUE PER 100 G	1 CUP = 104.0G	CORNMEAL, WHOLE-GRAIN, VALUE PER 100 G	1 CUP = 122.0G
Energy	kcal	340	408	404	420	362	442
Protein	g	13.21	15.85	14.66	15.25	8.12	9.91
Total lipid (fat)	g	2.5	3	9.12	9.48	3.59	4.38
Carbohydrate, by difference	g	71.97	86.36	65.7	68.33	76.89	93.81
Fiber, total dietary	g	10.7	12.8	6.5	6.8	7.3	8.9
Calcium, Ca	mg	34	41	55	57	6	7
Iron, Fe	mg	3.6	4.32	4	4.16	3.45	4.21
Magnesium, Mg	mg	137	164	144	150	127	155
Phosphorus, P	mg	357	428	452	470	241	294
Potassium, K	mg	363	436	371	386	287	350
Sodium, Na	mg	2	2	19	20	35	43
Zinc, Zn	mg	2.6	3.12	3.2	3.33	1.82	2.22
Vitamin C, total ascorbic acid	mg	0	0	0	0	0	0
Thiamin	mg	0.502	0.602	0.692	0.72	0.385	0.47
Riboflavin	mg	0.165	0.198	0.125	0.13	0.201	0.245
Niacin	mg	4.957	5.948	1.474	1.533	3.632	4.431
Vitamin B-6	mg	0.407	0.488	0.125	0.13	0.304	0.371
Folate, DFE	µg	44	53	32	33	25	30
Vitamin B-12	µg	0	0	0	0	0	0
Vitamin A, RAE	µg	0	0	0	0	11	13
Vitamin A, IU	IU	9	11	0	0	214	261
Vitamin E (alpha-tocopherol)	mg	0.71	0.85	0.7	0.73	0.42	0.51
Vitamin D (D2 + D3)	µg	0	0	0	0	0	0
Vitamin D	IU	0	0	0	0		
Vitamin K (phylloquinone)	µg	1.9	2.3	3.2	3.3	0.3	0.4

Source: USDA National Nutrient Database for Standard Reference 28 Software v.2.3.8

WILD RICE *and the* ENVIRONMENT

WILD RICE has great ecological value, and the streams and lakes where it grows abundantly support unusually diverse biological communities. Both migrating and resident wildlife rely on the nutritious and plentiful seed of natural wild rice. The dense stands provide cover for more than thirty species of waterfowl and nesting habitats for other bird species. Because the ripening of wild rice seeds coincides with fall migration, wild rice beds are one of the most important waterfowl foods in North America. The wood duck, canvasback, mallard, blue-winged teal, redhead, ring-necked duck, and other migratory waterfowl find stopover habitat among the plants. The rice is extremely important to the sora rail, a species of water bird that is in decline, providing up to 94 percent of its grains diet in the fall. The young shoots, germinating seeds, and mature stems and leaves also supply nourishment for wood ducks. According to ethnologist Albert Ernest Jenks, writing in 1901, many accounts by early Minnesota travelers describe "clouds of blackbirds [grackles], redwing blackbirds, and rice-birds [bobolinks] which subsist on the grain during and immediately after its milk stage."

Mammals such as the muskrat use the stalks for food. Wild rice provides food and shelter for many fish in North America, and the beds provide habitat for frogs and aquatic insects. The Comprehensive Wildlife Conservation Strategy of the Minnesota DNR lists seventeen species of wildlife known as "species in greatest conservation need" that use wild rice lakes as habitat for reproduction or foraging.

In addition, wild rice plants in our rivers and lakes minimize erosion by binding loose soils and acting as a buffer to slow winds across shallow wetlands. As water quality is stabilized, algal blooms are reduced and water becomes clearer. Annual reseeding ensures a stable ecosystem, protecting the other life-forms found in these water bodies.

Stands of wild rice have been gradually declining in Minnesota for many years. Rice can be damaged by pollution, large boat wakes, invasive species, and changes in water levels. Dams, the use of motorized boats, which can tear up the fragile stalks, runoff from nearby agriculture, and the introduction of exotic plants have all contributed to the decline.

Ojibwe bands have raised concerns about research conducted at the University of Minnesota's North Central Research and Outreach Center, specifically work in mapping the wild rice genome, which researchers began in 2003. The Ojibwe fear that this work will lead to genetic modifications of the plant and that the modified plant could contaminate the natural wild rice as wind or birds disperse its seeds. Several reservations have banned the introduction or growth of genetically modified wild rice seeds on their land. Presently there are no varieties of wild rice that are genetically modified, and none are expected in the foreseeable future, based on the limited acreage of wild rice.

Climate change is also likely to place stress on wild rice. Shifts in temperature, increases in pests and disease, changes in water level, and more frequent or more intense storms will have negative effects on both the natural stands and the cultivated beds. Climate change may also invite greater numbers of invasive plants that would harm the wild rice habitat.

DID YOU KNOW?

- The **SCIENTIFIC NAME** for wild rice is *Zizania palustris*; this species is an annual grass that has large, harvestable seeds. *Z. texana* is a perennial form that grows only in a small river in Texas; it has small seeds that are not harvested for food. *Z. aquatica* is an annual plant that grows in the St. Lawrence River and along the Atlantic and Gulf coasts of the United States; it has thin seeds and is rarely harvested for food.
- *Z. palustris* is found in the **FRESH WATER** of southern Canada, Michigan, Wisconsin, and Minnesota.
- The Ojibwe, Dakota, and other **GREAT LAKES TRIBES** use wild rice as an important staple in their diets.
- The French explorers called wild rice *foilles avoines*, translated as **"WILD OATS."**
- When cooked, wild rice **EXPANDS TO THREE OR FOUR TIMES** its original size.
- Wild rice is the only cereal grain **NATIVE TO NORTH AMERICA**.
- **OTHER NAMES** for wild rice are Canada rice, Indian rice, blackbird oats, false oats, American rice, and water oats. To the Dakota it is ***PSIN***.
- You can pop wild rice, like popcorn. Just heat it in a little oil and **SHAKE TILL IT POPS**.
- Lumberjacks ate **WILD RICE FOR BREAKFAST** with milk and honey.

SUMMARY AND CONCLUSION

MARY HOCKENBERRY MEYER

WHAT HAVE WE LEARNED?

We hope that exploring the 10 Plants presented in this book has been a fun and meaningful experience for you, the reader. We have attempted to highlight how these plants are unique to Minnesota's economy and to show the scale of their impact. Ideally, you will now look for and recognize the 10 Plants and think about how they affect your life. We hope you will recognize white pine on the horizon, whether in the suburbs, along the North Shore, or near Minnesota's many lakes, standing above the other trees as a supercanopy species. Look closely at the wetland you pass daily and watch the ebb and flow of purple loosestrife through the seasons and years. How diverse are the trees on your property? Are there any American elms? Which species could you add?

In surveying these ten plants, we have come to four conclusions: (1) species diversity is very important in our human-made landscapes; (2) natural plant communities are complex and intertwined as species have evolved together over time, along with human intervention and pressures; (3) history can be a great teacher if we pay attention; and (4) plants are essential for human health and well-being, providing us with great enjoyment, recreation, and of course food.

THE IMPORTANCE OF SPECIES DIVERSITY IN GROWING PLANTS

Monocultures, the planting of one species on a large scale, no matter how "good or resilient" that species is, can result in tragic devastation if the species is exposed to a new disease or insect. The American elm is a tough, fast-growing, resilient native species, but it succumbed rapidly to Dutch elm disease. Luckily for Minnesota, we learned from the devastation of the elms in the eastern United States and could plan for and manage the disease by the time it moved to the Midwest. Although we lost (and continue to lose) many trees, we still have some streets with stately, healthy elms. We quickly learned the importance of planting diverse species in urban areas. The loss of many American elms has taught us to plant a wide range of species, both native and nonnative, in our communities to withstand the pests and extreme weather of the twenty-first century. The city of Plymouth, for example, is planting sixteen kinds of trees to replace the ash it is removing due to the newest pest: the emerald ash borer.

Plants growing in a diverse community of species, as they do in nature, prevent the development of large populations of insects or diseases. When we grow plants in large artificial monocultures, as has been done with corn, wheat, soybeans, and American elm street trees, we create a huge area of uniformity that can result in high pest populations. Diversity is an important concept that, when overlooked, as we see with wheat, purple loosestrife, and the American elm, can lead to major losses of income (wheat rust), habitat (invasive plants), and aesthetics (street trees). Growing a variety of plants, including different genotypes within a species, results in less loss to disease, more genes for building resistance, and a better chance for adaptation to diverse growing conditions. Species diversity not only slows the development of pests but also provides niches for a wide variety of plants to coexist. Many of the 10 Plants are grown as monocultures; this is true not just for soybeans and corn but also for turfgrass, street tree selection, and even food crops, such as apples and wild rice. Although monocultures are good for high yields, they must be carefully managed for pests and nutrient depletion or runoff. Hopefully we have learned that monocultures of all types—in wetlands, on boulevards, or in farm fields—can have adverse environmental impacts and can be very costly to remedy.

COMPLEXITY OF NATURAL SYSTEMS

Plants provide habitat and food sources for insects, birds, and animals. Depleting white pine changed the species composition not only of plants in northern Minnesota but also of insects, birds, and animals that lived in white pine forests. Fewer white pines meant fewer supercanopy nesting sites for bald eagles, less shelter for black bears, and reduced numbers of nesting sites for small mammals. Changing one plant's population can have a ripple effect whose results may not be apparent for years. When this shift happens over generations, as it did with white pine, it becomes harder to recognize or recall.

We know today that wild rice responds to changes in water levels as well as the nutrients in the water. Mining, farming, and manufacturing miles away from a lake can still affect wild rice growth and persistence over time. Rarely if ever is one environmental change isolated from other parts of the same or nearby ecosystems.

Complex systems do not have simple answers for what is best or right, especially when economic changes (such as new businesses that create jobs and steady employment in rural regions) must be balanced with their environmental effects on groundwater, soil, lakes, and natural resources. These decisions we face are very difficult and require thoughtful study, research, education, and public input. They will become even more challenging as our population grows and uses more resources.

HISTORIC CHOICES AND FUTURE CHOICES

Lessons from the past may be the hardest thing to learn, because it seems we have such short memories, and who carries the memories from one generation to the next? No one is alive today who can talk about their experiences logging the white pine from northern Minnesota or shipping Minnesota wheat to Europe, making the state the milling capital of the world. We read about it and wonder what it was like to be there in person. What did the woods look like before white pine was harvested? When changes happen over more than one generation, we tend to forget the consequences. But faster changes, like those that resulted from Dutch elm disease, are more easily remembered; consequently, no city park departments are planting monocultures of street trees today. Things that happen within one lifetime, such as the destruction caused by Dutch elm disease, thus are more likely to alter our behavior.

We know that some plants can become widespread due to self-seeding; we remember the before and after of purple loosestrife. History can be an effective teacher—if we consult it for help in making our choices today.

APPRECIATING PLANTS AND UNDERSTANDING THEIR VALUE

We hope readers will see the 10 Plants "with new eyes," as Marcel Proust said, and understand the complexity of natural ecosystems and how they interact with the state's economy. Too often we take plants for granted; we fail to notice what is right in front of us. We pay attention to animals and birds and may trek miles to watch their actions, but we often fail to notice which plants these birds and animals are nesting in or feeding on. What will wake us up so that we can see plants and value them? Environmentalist Paul Hawken has said, "The average citizen can recognize 1,000 brand names and logos but fewer than 10 local plants." How can we value that which we cannot name? How can we realize that lost plants, such as trees that are hundreds of years old, take with them much more than meets the eye?

THE 10 PLANTS OF 2100

If the 10 Plants were chosen in 1900, would they be the same as the ones selected in 2012? Wheat, white pine, and maybe corn would make that list, but probably not apples or turfgrass and certainly not purple loosestrife or soybeans. What plants would make the list in 2100? If history is our teacher, there will be plants we do not know or cultivate today—possibly a new blueberry or other food probably grown for health benefits. Perhaps the list will include amaranth (today considered a weed in Minnesota, but grown as a food crop for thousands of years in Africa); it is high in iron, zinc, protein, calcium, and magnesium and is gluten-free. Perhaps a new crop for biomass or producing fuel or even poultry bedding? Perhaps by 2100 a new plant grown specifically for pollinators to support native bees will be on the list of plants changing Minnesota.

With approximately 400,000 known species of plants, it is rather amazing that we can identify these ten that have made such an impact on Minnesota. Humans and plants will continue to evolve in a complex relationship. We hope this book has shown the historical significance of the 10 Plants and the importance of plant diversity, and that it will help readers see and appreciate plants and begin to grasp the complexity of natural systems.

The story of the 10 Plants is about choices we have made, some for economics (corn and soybeans), some for the environment (purple loosestrife), some for social acceptance (lawns). We hope this book will help readers appreciate not only the 10 Plants that changed Minnesota but all the plants they pass daily growing in fields, parks, gardens, or forests.

ACKNOWLEDGMENTS

We are very grateful to many people who assisted in making this book a reality.

Thanks to the 2012 10 Plants judges: Alan Ek, professor, University of Minnesota Department of Forest Resources; Al Withers, Director, Minnesota Agriculture in the Classroom; Beverly Durgan, Dean of University of Minnesota Extension; Bob Quist, Director, Oliver Kelley Farm, Minnesota Historical Society; Brian Buhr, Dean, College of Food, Agricultural and Natural Resource Sciences; Karen Kaler of the University of Minnesota's President's Office; Mary Maguire Lerman, President, Minnesota State Horticultural Society; Nancy Jo Ehlke, professor, University of Minnesota Department of Agronomy and Plant Genetics; Susan Bachman West, Owner, Bachman's, Inc.; Gary Gardner, professor, University of Minnesota Department of Horticultural Science; Neil Anderson, professor, University of Minnesota Department of Horticultural Science; Karl Foord, University of Minnesota Extension Educator.

Thanks to staff and colleagues at the Minnesota Landscape Arboretum and the Andersen Horticultural Library: Katherine Allen, Judy Hohmann, Barb DeGroot, Peter Moe, Frank Molek, Leslie Nitabach, Tim Kenny, Sandy Tanck, Randy Gage, David Matteson.

Special thanks to chapter reviewers and experts who assisted in locating information: Karl Mueller, Phillip Potiand, Gary Johnson, Chad Giblin, Garrett Beier, Brad Carlson, Mike Schmitt, Emily Hoover, David Bedford, Jim Luby, Luke Skinner, Richard Rezanka, Jerry Fruin, Jim Orf, Michael Russelle, Erv Oelke, Sam Bauer, Craig Sheaffer, James A. Anderson, Beth Nelson of the Wild Rice Council, Bill Lazarus, Jeff Sauve of the St. Olaf Library, Keah Kodner of the James J. Hill Library, Dan Braaten of the North Central Research Extension Center, Kevin Morris of the National Turfgrass Association,

Katy Goetz, Kurtis Greenley, John Carstad, Doug Tallamy, Kimberley Shropshire, and many others.

Thanks to Arne Carlson for his insightful foreword and to the editors at Minnesota Historical Society Press: Shannon Pennefeather, Pamela McClanahan, and assistants Katy Goetz and David Katz.

Special thanks to University of Minnesota photographer David L. Hansen, for his many images and his willingness to share them; and to the late Cathy Gilchrist Kaudy, who submitted lovely watercolor illustrations of her nominations for the 10 Plants and drew the final selections for use on the website and in this book. Thanks to Erich Kaudy for allowing us to use Cathy's watercolors.

Thanks to our families, especially Jim Meyer for his patience and support.

MARY H. MEYER *and* SUSAN D. PRICE

SOURCES

INTRODUCTION

The Minnesota Historical Society owns the original F. J. Marschner Map; the five-foot-tall map is large enough to locate specific areas within a county and is accessible online: http://discussions.mnhs.org/collections/2008/09/marschner-map-of-original-vegetation/.

"Daily Media Use among Children and Teens Up Dramatically from Five Years Ago." Kaiser Family Foundation. January 20, 2010. http://kff.org/disparities-policy/press-release/daily-media-use-among-children-and-teens-up-dramatically-from-five-years-ago/.

"How Does Nature Impact Our Wellbeing?" Center for Spirituality and Healing, University of Minnesota. http://www.takingcharge.csh.umn.edu/enhance-your-wellbeing/environment/nature-and-us/how-does-nature-impact-our-wellbeing. Accessed February 9, 2016.

Wandersee, J., and E. Schussler. "Toward a Theory of Plant Blindness." *Plant Science Bulletin* 47 (2001): 2–9. http://www.botany.org/bsa/psb/2001/psb47-1.html#Toward a Theory of Plant.

1. ALFALFA

"Alfalfa: Energy Economy Environment." Alfalfa Council. http://www.alfalfa.org/pdf/AlfalfaEnergyEconEnv.pdf. Accessed November 16, 2016.

Balliette, John, and Ron Torell. "Alfalfa for Beef Cows." University of Nevada Reno Cooperative Extension. http://www.unce.unr.edu/publications/files/ag/other/fs9323.pdf. Accessed November 16, 2016.

"The Big Show at Ortonville." *Willmar Tribune*, September 28, 1915, 4.

Brand, Charles J. "Grimm Alfalfa and Its Utilization in the Northwest." *Bulletin of Plant Industry* 209 (1911): 14.

Brough, R. Clayton, Lauren R. Robinson, and Richard H. Jackson. "The Historical Diffusion of Alfalfa." *Journal of Agronomic Education* 6 (1977): 13–19.

Carlsted, John. Personal communication with Mary Meyer. July 10, 2015.

Cook, Rob. "Alfalfa Production by State (2012 vs. 2013)." December 20, 2014. http://beef2live.com/story-alfalfa-production-state-2012-vs-2013-0-107297.

Coulter, Jeff, Michael Russelle, Craig Sheaffer, and Dan Kaiser. "Maximizing On-Farm Nitrogen and Carbon Credits from Alfalfa to Corn." Interim Technical Report for Period of April 15, 2009–December 31, 2009. Research project for the Minnesota Corn Research and Promotion Council.

Cumo, Christopher. "Alfalfa." In *Encyclopedia of Cultivated Plants, from Acacia to Zinnia*, 9–15. Santa Barbara, CA: ABC-CLIO, 2013.

"Don't You Want to Come in for Alfalfa Seed at Cost?" *Willmar Tribune*, December 17, 1913, 1.

"Editors Can't Stay Away." *Willmar Tribune*, December 3, 1913, 10.

Edwards, Everett E. "Wendelin Grimm and Alfalfa." *Minnesota History* 19.1 (1938): 21–33. http://collections.mnhs.org/MNHistoryMagazine/articles/19/v19i01p021-033.pdf.

Food for Life. "Forage Legumes." University of Minnesota, Minnesota Agricultural Experiment Station, 2001, 3–4.

Gould, Heidi. "Grimm, Wendelin (1818–1890) and 'Grimm Alfalfa.'" MNopedia. November 17, 2015. http://www.mnopedia.org/person/grimm-wendelin-1818-1890-and-grimm-alfalfa.

Hannaway, David B., and Christina Larson. "Alfalfa (*Medicago sativa L.*)." Forage Information System, Oregon State University. 2004.

Hirata, Masahiko. "Forage Crop Production." In *The Role of Food, Agriculture, Forestry and Fisheries in Human Nutrition*. Vol. 1, *Encyclopedia of Life Support Sys-*

tems. Oxford: Eolss Publishers, 2011. http://www.eolss.net/sample-chapters/c10/e5-01a-01-08.pdf.

Johnson, Andrea. "Research Continues on Value of N from Alfalfa Ahead of Corn." *Minnesota Farm Guide*. 2014. http://www.minnesotafarmguide.com/news/crop/research-continues-on-value-of-n-from-alfalfa-ahead-of/article_bd8382fc-ba7c-11e3-9428-0019bb2963f4.html.

Jung, Dr. Hans. "Alfalfa: A Sustainable Crop for Biomass Energy Production." http://www.ars.usda.gov/SP2UserFiles/Place/36401000/AlfalfaforBiomass.pdf. Accessed November 16, 2016.

Kelzer, Frank. "The History of Grimm Alfalfa." University of Minnesota Extension. 1957. http://www.extension.umn.edu/agriculture/forages/variety-selection-and-genetics/docs/ext-umn-history-of-grimm-alfalfa-1957.pdf.

"Many Prizes at Morris Show." *Willmar Tribune*, October 29, 1913, 5.

National Alfalfa and Forage Alliance. "Mission and Objective." 2016. www.alfalfa.org/.

National Register of Historic Places. "Grimm, Wendelin, Farmstead, Victoria, MN." March 4, 2012. http://www.waymarking.com/waymarks/WMDX2M_Grimm_Wendelin_Farmstead_Victoria_MN.

"'On to Benson' Is the Slogan." *Willmar Tribune*, November 18, 1914, 7.

Putnam, Dan, Michael Russelle, Steve Orloff, Jim Kuhn, Lee Fitzhugh, Larry Godfrey, Aaron Kiess, Rachael Long. "Alfalfa, Wildlife and the Environment: The Importance and Benefit of Alfalfa in the 21st Century." California Alfalfa and Forage Association. 2001. http://alfalfa.ucdavis.edu/-files/pdf/Alf_Wild_Env_BrochureFINAL.pdf.

"Red River Valley's Week Made New Record." *Northwest Monthly* 10.3 (February 1926): 1.

Riemers, Richard. "Cover Story: The Legacy of Wendelin Grimm and His Everlasting Clover." *The Land* (Mankato, MN), February 10, 2010. http://www.thelandonline.com/l_history/x1004922965/Cover-story-The-legacy-of-Wendelin-Grimm-and-his-everlasting-clover.

Russelle, Michael. "Alfalfa." *American Scientist*, May–June 2001. http://www.americanscientist.org/issues/num2/alfalfa/1.

"Spirit of Festival Will Pervade All." *Willmar Tribune*, December 3, 1913, 10.

Summers, Charles, Larry Godfrey, and Eric Natwick. "Managing Insects in Alfalfa." University of California, Davis, Division of Agriculture and Natural Resources. December 2007. http://alfalfa.ucdavis.edu/IrrigatedAlfalfa/pdfs/UCAlfalfa8295Insects.free.pdf.

"Town and Country Observe 'Alfalfa Day.'" *The Banker-Farmer*, February, 1925, 15.

USDA, National Agricultural Statistics Service. "Alfalfa Hay, Acreage Harvested, Yield." Crop county estimates. https://www.nass.usda.gov/Statistics_by_State/Minnesota/Publications/County_Estimates/index.php. Accessed November 16, 2016.

———. "Minnesota Ag News: Crop Production." 2014. https://www.nass.usda.gov/Statistics_by_State/Minnesota/Publications/Crops_Press_Releases/2015/MN_CropProd_11_15.pdf.

———. "Minnesota Agricultural Statistics." 2013–14. http://www.nass.usda.gov/Statistics_by_State/Minnesota/.

"Wendelin Grimm Farmstead." Wikipedia. September 12, 2016. https://en.wikipedia.org/wiki/Wendelin_Grimm_Farmstead.

"Workshop Touts Potential for Alfalfa as Cellulosic Feedstock." *High Plains/Midwest Ag Journal*. July 2, 2010. http://www.hpj.com/archives/workshop-touts-potential-for-alfalfa-as-cellulosic-feedstock/article_5f51fb88-e8cf-5957-8e46-416031dc42a2.html.

ALFALFA: ROW AND PERENNIAL CROP COMPARISON TABLE (PAGE 27)

"Before You Start an Apple Orchard." University of Minnesota Extension. 2016. fruit.cfans.umn.edu/apples/beforeyoustart/.

University of Minnesota, Center for Farm Financial Management. Finbin. 2016. https://finbin.umn.edu.

USDA, National Agriculture Statistics Service. "Statistics by State: Minnesota." 2015. https://www.nass.usda.gov/Statistics_by_State/Minnesota/Publications/Annual_Statistical_Bulletin/2015/MN%20Bulletin%202015_Page11.pdf.

2. AMERICAN ELM

American Elm Project. 2010. http://americanelmproject.org/. Accessed 2014.

"American Elms." Growing History: The Philadelphia Historic Plants Consortium. March 31, 2012. https://growinghistory.wordpress.com/2012/03/31/american-elms/.

Beaulieu, David. "Dutch Elm Disease on American Elm Trees." About Home. http://landscaping.about.com/cs/treesshrubs/a/american_elms.htm. Accessed November 16, 2016.

"Benefits of Trees." International Society of Arboriculture. 2011. http://www.treesaregood.org/treeowner/benefitsoftrees.aspx.

"Benefits of Trees and Urban Forests." ACTrees/Alliance for Community Trees. August 2011. actrees.org/files/Research/benefits_of_trees.pdf.

Campanella, Thomas J. *Republic of Shade: New England and the American Elm*. New Haven, CT: Yale University Press, 2003.

Downing, Andrew Jackson. *A Treatise on the Theory and Practice of Landscape Gardening, Adapted to North America*. New York: Wiley and Putnam, 1841.

French, David W. "A History of Dutch Elm Disease in Minnesota." University of Minnesota Extension. 1993.

Giblin, Chad P. "Elms for Minnesota Home Landscapes." *Yard & Garden News*, July 1, 2007.

Giblin, Chad P., research fellow, University of Minnesota. Phone conversation. January 15, 2013.

Grabowski, Michelle. "Dutch Elm Disease—Past, Present, and Future." *Yard & Garden News*, July 1, 2007.

"Minneapolis History. Cities of the United States." Encyclopedia.com. 2006. http://www.encyclopedia.com/topic/Minneapolis.aspx.

Moran, Mark. "American Elm, Relationships in Nature." http://www.fcps.edu/islandcreekes/ecology/american_elm.htm. Accessed November 16, 2016.

Mueller, Karl, City of St. Paul forester. E-mail conversations. June 2015.

Potiand, Phillip, Minneapolis Parks and Recreation forester. Phone conversation. April 15, 2015.

Rutkow, Eric. *American Canopy: Trees, Forests, and the Making of a Nation*. New York: Scribner, 2013.

Stennes, Mark. "The Mystery of the Elms in the Kandiyohi Forest." *Minnesota Native Plant Society Newsletter* 27.4 (Summer 2008): 1–3.

"*Ulmus Americana*." Wikipedia. October 23, 2016. http://en.wikipedia.org/wiki/Ulmus_americana.

"*Ulmus Americana*, American Elm." *Midwest Gardening*. http://midwestgardentips.com/american_elm.html. Accessed November 16, 2016.

United States National Arboretum. "*Ulmus americana:* 'Valley Forge' and 'New Harmony.'" March 31, 2011. http://www.usna.usda.gov/Newintro/american.html.

Wirth, Theodore. "Memorial Trees: The Victory Drive of Minneapolis, Minnesota." *Parks and Recreation Magazine* 3 (January 1920): 21.

Writing the Story of Saint Paul's Historic Elms. St. Paul Parks and Recreation, the University of Minnesota, St. Paul's Tree Advisory Panel, University of St. Thomas Environmental Studies Field Seminar Class. 2014.

3. APPLES

"75 Years of Producing Hardy Fruits for the North." *Minnesota Horticulturist* (October/November 1983): 238–42.

"Apples." University of Minnesota. 2012. http://www.apples.umn.edu/.

"Apples and More: Apple Facts." University of Illinois Extension. 2008. http://urbanext.illinois.edu/apples/facts.cfm.

Bedford, David, senior research fellow, University of Minnesota. Phone conversation. January 10, 2012.

Curtiss-Wedge, Franklin. *History of Freeborn County, Minnesota*. Chicago: H. C. Cooper, Jr. and Co., 1911.

"The Father of Orchardists." *Minnesota Calls* (August 1990): 39.

Haralson, Charles. "How Members May Assist the State Fruit Breeding Farm." *Minnesota Horticulturist* (November 1908): 401.

———. "The State Fruit Breeding Farm in 1922." *Minnesota Horticulturist* (January 1922): 34.

"History of Fruit Research." Minnesota Landscape Arboretum. August 2007. http://www.arboretum.umn.edu/fruitbreeding.aspx.

Hoover, Emily, David Bedford, and Doug Foulk. "Apples for Minnesota and Their Culinary Uses." University of Minnesota Extension. 2009. http://www.extension.umn.edu/garden/yard-garden/fruit/apples-for-minnesota-and-their-culinary-uses/.

Hutchins, A. E. "Historical Notes." *Minnesota Horticulturist* (June 1966): 73.

Kusten, Mary Ellen. "An Apple a Day, Too Much Pesticide Spray." *AGMag*, February 26, 2015. http://www.ewg.org/agmag/2015/02/apple-day-too-much-pesticide-spray.

Luby, Dr. James, University of Minnesota professor. Phone conversation. April 2, 2015.

"Minnesota Apples." Minnesota Apple Growers Association. 2012. http://www.minnesotaapple.org/.

"Minnesota Orchards: Minnesota's Distinctive Apples." *Apple Journal* (2001). http://www.applejournal.com/mn01.htm.

"Peter Gideon Farmhouse." Wikipedia. http://en.wikipedia.org/wiki/Peter_Gideon_Farmhouse. Accessed June 2016.

Pliny the Elder. *Naturalis Historia* (*Natural History*). Libri Vox. https://librivox.org/author/5293?primary_key=5293&search_category=author&search_page=1&search_form=get_results. Accessed November 16, 2016.

Pollan, Michael. *The Botany of Desire: A Plant's-Eye View of the World*. New York: Random House, 2001.

Qualey, Carlton C. "Diary of a Swedish Immigrant Horticulturist, 1855–1898." *Minnesota History* 43.2 (Summer 1972): 63–70. http://collections.mnhs.org/MNHistoryMagazine/articles/43/v43i02p063-070.pdf.

Thompson, Ruth. "Ye Old Time Minneapolis Gardens." *Hennepin County History* (April 1942): 5.

US Apple Association. "All about Apples and Apple Products." 2012. http://www.usapple.org/.

———. "Apple Health Benefits." February 2016. http://usapple.org/wp-content/uploads/2016/04/Health-Benefits-Research-Summary_Feb2016-2.pdf.

US Forest Service. "Minneapolis' Urban Forest." http://www.na.fs.fed.us/urban/treespayusback/vol1/ufore%20mpls%20summary.pdf. Accessed November 16, 2016.

USDA, National Agricultural Statistics Service. "2013 Census of Agriculture." Minnesota. http://www.nass.usda.gov/Statistics_by_State/Minnesota/. Accessed November 16, 2016.

Warshaw, Hope. "Apples: To Peel or Not To Peel?" *Washington Post*, December 2, 2014. https://www.washingtonpost.com/lifestyle/wellness/apples-to-peel-or-not-to-peel/2014/12/01/f9f97e9e-74d5-11e4-9d9b-86d397daad27_story.html.

4. CORN

Allen, Brad. "Minnesota Corn Belt Weathers Today's Floods and Offers Rich History." *MinnPost*, October 8, 2010. http://www.minnpost.com/business/2010/10/minnesota-corn-belt-weathers-todays-floods-and-offers-rich-history.

Balmer, Frank E. "The Farmer and Minnesota History." *Minnesota History* 7.3 (1926): 199–217. http://collections.mnhs.org/MNHistoryMagazine/articles/7/v07i03p199-217.pdf.

Burris, Evadene A. "Frontier Food." *Minnesota History* 14.4 (December 1933): 378–92. http://collections.mnhs.org/MNHistoryMagazine/articles/14/v14i04p378-392.pdf.

Carroll, Sean B. "Tracking the Ancestry of Corn Back 9,000 Years." *New York Times*, May 24, 2010. http://www.nytimes.com/2010/05/25/science/25creature.html?_r=0.

"Corn. Rooted in Human History." National Corn Growers Association. 2012. www.ncga.com/file/574.

"Corn Boom Could Expand 'Dead Zone' in Gulf." NBC News. December 17, 2007. http://www.nbcnews.com/id/22301669/.

da Fonseca, Rute R., et al. "The Origin and Evolution of Maize in the Southwestern United States." *Nature Plants* 1 (2015), article 14003. http://www.nature.com/articles/nplants20143.

Deiss, Ron. "Corn Husking Once Excited the Nation." The Farmer.com. March 2009. http://magissues.farmprogress.com/tfm/TF03Mar09/tfm029.pdf.

Fitzgerald, Emma. "Land of 10,000 Polluted Lakes." Minnesota 20/20. June 2, 2014. http://www.mn2020.org/issues-that-matter/energy-environment/land-of-10000-polluted-lakes.

Foley, Jonathan. "It's Time to Rethink America's Corn System." *Scientific American*, March 5, 2013. http:

//www.scientificamerican.com/article/time-to-rethink-corn/.

Fussell, Betty. *The Story of Corn*. New York: North Point Press, Farrar, Straus and Giroux, 1992.

Granger, Susan, and Scott Kelly. "Historic Context, Study of Minnesota Farms, 1820–1960." Gemini Research, prepared for the Minnesota Department of Transportation. 2005. www.dot.state.mn.us/culturalresources/farmsteads.html.

"Growing Corn the 'Original' Way." Science Buzz. 2004–16. www.sciencebuzz.org/topics/growing-corn-original-way.

Hakin, Danny. "Doubts about a Promised Bounty." *New York Times*, October 29, 2016. http://www.nytimes.com/2016/10/30/business/gmo-promise-falls-short.html?_r=0, accessed December 15, 2016.

Hardeman, Nicholas P. *Shucks, Shocks, and Hominy Blocks: Corn as a Way of Life in Pioneer America*. Baton Rouge: Louisiana State University Press, 1981.

Hirst, K. Kris. "The Domestication of Maize." About-Education. 2016. http://archaeology.about.com/od/mterms/qt/maize.htm.

"Historic Context Study of Minnesota Farms, 1820–1960." Minnesota Department of Transportation. 2015. http://www.dot.state.mn.us/culturalresources/farmsteads.html.

Jarchow, Merrill D. *The Earth Brought Forth: A History of Minnesota Agriculture to 1885*. St. Paul: Minnesota Historical Society, 1949.

Johnson, Andrew. "Small Seed Corn Companies Offer Quality Products Too." *Minnesota Farm*. January 29, 2014. http://www.minnesotafarmguide.com/feature/seed_guide/small-seed-corn-companies-offer-quality-products-too/article_3830fe6a-8b8b-11e3-8aaa-0019bb2963f4.html.

Larson, W. E., and V. B Cardwell. "History of U.S. Corn Production." University of Minnesota. 1999. http://www2.econ.iastate.edu/classes/econ496/lence/spring2004/corn.pdf.

Lavenda, Robert H. *Corn Fests and Water Carnivals: Celebrating Community in Minnesota*. Washington, DC: Smithsonian Institute Press, 1997.

Letterman, Edward J. *Farming in Minnesota*. St. Paul, MN: Ramsey County Historical Society, 1961.

Marcotty, Josephine. "Four Minnesota Watersheds Will Be Test of Ag Cleanup Plan." *Star Tribune*, June 11, 2013. http://www.startribune.com/four-minnesota-rivers-will-test-9–5-million-agriculture-cleanup-plan/210944091/.

———. "Nitrogen Pollution Widespread in Southern Minnesota Waters, Report Finds." *Star Tribune*, June 27, 2013. http://www.startribune.com/nitrogen-pollution-widespread-in-southern-minnesota-waters-report-finds/213247091/.

Maxwell, Davis. "Corn Domesticated from Mexican Wild Grass 8,700 Years Ago." *National Geographic*, March 23, 2009. http://voices.nationalgeographic.com/2009/03/23/corn_domesticated_8700_years_ago/.

"Me So Corny: Big Corn of Minnesota." Highway Highlights. January 30, 2014. http://highwayhighlights.com/2014/01/me-so-corny-big-corn-of-minnesota/.

"Minnesota 13." Wikipedia. http://en.wikipedia.org/wiki/Minnesota_13. Accessed November 16, 2016.

"Minnesota Corn Farmers Continue to Drive Minnesota's Economy." Minnesota Corn Growers Association. February 19, 2013. http://mncorn.org/media-center/daily-story/minnesota-corn-farmers-continue-drive-minnesotas-economy.

"The Minnesota Nutrient Reduction Strategy." Minnesota Pollution Control Agency. 2014. https://www.pca.state.mn.us/water/nutrient-reduction-strategy#nutrient-strategy-1920a62e.

National Academies of Sciences, Engineering, and Medicine. *Genetically Engineered Crops: Experiences and Prospects*. Washington, DC: The National Academies Press, 2016. doi: 10.17226/23395.

Philpott, Tom. "King Corn Mowed Down 2 Million Acres of Grassland in 5 Years Flat." *Mother Jones*, February 20, 2013. http://www.motherjones.com/tom-philpott/2013/02/king-corn-gobbles-climate-stabilizing-grassland-midwest.

Schoolcraft, Henry Rowe. *Narrative Journals of Travels through the Northwestern Regions of the United States*. Albany, NY: E & E Hosford, 1821.

"Sweet Corn." Agricultural Marketing Resource Center. May 2015. http://www.agmrc.org/commodities-products/vegetables/sweet-corn/.

Tvetden, Lenny. "Martin County Event Attracts 50,000 to 100,000 People." Martin County Historical Society. http://fairmont.org/mchs/National%20Corn%20

Husking%20Contest.pdf. Accessed November 16, 2016.
Weber, Eric W. "Peas, Corn and Beyond: Minnesota's Green Giant Company Was a Canned Food Pioneer." *MinnPost*, January 8, 2013. https://www.minnpost.com/mnopedia/2013/01/peas-corn-and-beyond-minnesotas-green-giant-company-was-canned-food-pioneer.
Wilde, Amy. "The Revolution of Seed Corn." *Litchfield Independent Review*, business section, November 10, 2013. http://www.crowrivermedia.com/independentreview/news/business/the-revolution-of-seed-corn/article_13e4d2b6-7446-5ab5-97f1-2e16d28d8460.html.
Winchell, N. H., Jacob V. Brower, Alfred J. Hill, and Theodore H. Lewis. *The Aborigines of Minnesota: A Report Based on the Collections of Jacob V. Brower, and on the Field Surveys and Notes of Alfred J. Hill and Theodore H. Lewis*. St. Paul: Minnesota Historical Society, 1911.
USDA, Minnesota Department of Agriculture. "Minnesota Agricultural Statistics." 2012. http://www.nass.usda.gov/Statistics_by_State/Minnesota/.
USDA, National Agricultural Statistics Service, Minnesota Field Office. "Minnesota Ag News—Corn County Estimates." 2014. https://www.nass.usda.gov/Statistics_by_State/Minnesota/Publications/County_Estimates/.

5. LAWNS AND TURF

Andrews Nursery Company. "Outdoor Living Rooms." Faribault. 1936.
Appleton, Jay. *The Experience of Landscape*. New York: Wiley and Sons, 1975.
Beard, J. B., and R. L. Green. "The Role of Turfgrasses in Environmental Protection and Their Benefits to Humans." *Journal of Environmental Quality* 23 (1994): 252–64.
"Better Fields for Better Play." *Baseball Field Maintenance Handbook*. 2015. http://www.ultimate-baseball-field-renovation-guide.com/baseball-field-maintenance-handbook.html?hop=turnkey21.
Brown, George E., III. *100 Years of Minnesota Golf*. Edina: Minnesota Golf Association, 2001.
Butzer, Karl W. "Environment, Culture, and Human Evolution." *American Scientist* 65.5 (September–October 1977): 572–84.
Carey, Richard O., et al. "A Review of Turfgrass Fertilizer Management Practices: Implications for Urban Water Quality." *HortTechnology* 22 (2012): 280–91.
Diep, Francie. "Lawns vs. Crops in the Continental U.S." ScienceLine. July 3, 2011. http://scienceline.org/2011/07/lawns-vs-crops-in-the-continental-u-s/.
"Fruit Growers near Prairieville." *The Minnesota Monthly* (September 1869): 321.
Hard, C. Gustav. *Landscaping Your Home*. University of Minnesota Extension Bulletin. 1958.
Hayden, Thomas. "Could the Grass Be Greener?" *U.S. News & World Report*, May 8, 2005.
Haydu, John J., Alan W. Hodges, and Charles R Hall. "Economic Impacts of the Turfgrass and Lawncare Industry in the United States." University of Florida, IFAS Extension. 2006. http://edis.ifas.ufl.edu/fe632.
"The History of Lawns in America." American Lawns. 1997. http://www.american-lawns.com/history/history_lawn.html.
"How Miracle Green Treated Lawn Seed." *Milwaukee Sentinel*, May 3, 1953, 4.
Jenkins, Virginia Scott. *The Lawn: A History of an American Obsession*. Washington, DC: Smithsonian Institution Press, 1994.
Kaiser, Laura Fisher. "Low-Maintenance Lawn Alternatives: Ground Cover." House Logic. 2015. http://www.houselogic.com/home-advice/lawns/low-maintenance-lawn-alternatives-ground-cover/.
The Lawn Institute. "Interesting Facts about Turfgrass." Professional Lawn Care Association of America. N.d.
Lavidis, Nick. "Does Gardening Improve Your Memory?" UQ News. August 20, 2009. University of Queensland. https://www.uq.edu.au/news/article/2009/08/does-gardening-improve-your-memory.
Longley, Dr. L. E. "Care of Lawns in Spring." *Minnesota Horticulturist* (April 1938).
———. "Crabgrass Control." *Minnesota Horticulturist* (June 1940).
Meyer, Mary Hockenberry, Bridget K. Behe, and James Heilig. "The Economic Impact and Perceived Environmental Effect of Home Lawns in Minnesota." *HortTechnology* 11.4 (October–December 2001): 585–90.
Morris, Kevin R. "The National Turfgrass Research Ini-

tiative." August 1, 2006. http://gsrpdf.lib.msu.edu/ticpdf.py?file=/2000s/2006/060926.pdf.

Northrup King Company Records. Manuscript collection. Minnesota Historical Society, St. Paul.

Perry, Dr. Leonard. "Fuel-Efficient Lawn and Landscape." University of Vermont Extension, Department of Plant and Soil Science. http://pss.uvm.edu/ppp/articles/fuels.html. Accessed November 16, 2016.

Ratliff, Evan. "Turf Wars: The Battle over the American Lawn." *Atavist Magazine*. http://www.atavistic.org/evan/images/ReadyMade/Lawn%20Wars.pdf. Accessed November 16, 2016.

Robbins, Paul. *Lawn People: How Grasses, Weeds, and Chemicals Make Us Who We Are*. Philadelphia: Temple University Press, 2007.

Rooney, Theresa. "The American Lawn: A Brief History." Do It Green! Minnesota. November 1, 2009. http://www.doitgreen.org/green-living/american-lawn-brief-history.

Science of (the) Green. University of Minnesota. 2015. http://recwell.umn.edu/golf/science-of-the-green.

Scott, Frank J. *The Art of Beautifying Suburban Home Grounds of Small Extent*. New York: D. Appleton and Company, 1870.

Shefchik, Rick. *From Fields to Fairways: Classic Golf Clubs of Minnesota*. Minneapolis: University of Minnesota Press, 2012.

Snyder, Leon C. "The Home Lawn." *Minnesota Horticulturist* (April 1953): 59.

Stadtherr, Richard J. "Better Lawns—Through Research." *Minnesota Horticulturist* (April 1956): 36.

———. "Don't Neglect Your Lawn; It is the Most Important Feature of the Landscape." *Minnesota Horticulturist* (September 1956): 99, 111.

Steinberg, Ted. *American Green: The Obsessive Quest for the Perfect Lawn*. New York: Norton, 2006.

SULIS: Sustainable Urban Landscape Information Series. "Environmental Benefits of a Healthy, Sustainable Lawn." University of Minnesota. 2006. http://www.sustland.umn.edu/maint/benefits.htm.

Thesier, Kelly. "With Grass, Target Field Is a Reality." MLB.com. August 25, 2009. https://www.google.com/#q=thesier%2C+Kelly%2C+with+grass+target+field.

"Think Clean Air Landscaping." Hennepin County Environmental Services. http://www.hennepin.us/~/media/hennepinus/residents/environment/documents/landscaping-guide-ch-11-resources.pdf?la=en. Accessed November 16, 2016.

"Turf Grass Madness: Reasons to Reduce the Lawn in Your Landscape." University of Delaware Cooperative Extension. March 10, 2010. http://ag.udel.edu/udbg/sl/vegetation/Turf_Grass_Madness.pdf.

6. PURPLE LOOSESTRIFE

Anderson, Neil. "Purple Loosestrife." 10 Plants That Changed Minnesota. Lecture. Minnesota Landscape Arboretum, Chanhassen, MN. September 20, 2012.

Anderson, Neil O., and Peter D. Ascher. "Fecundity and Fitness in Cross-Compatible Pollinations of Tristylous North American *Lythrum salicaria* Populations." *Theoretical and Applied Genetics* 101 (2000): 830–43.

Begg, Virginia Lopez. "Pros and Cons of a Well Known Perennial." *New York Times*, August 18, 1985. http://www.nytimes.com/1985/08/18/arts/pros-and-cons-of-a-well-known-perennial.html.

"Biological Control of Purple Loosestrife." Minnesota Department of Natural Resources. http://www.dnr.state.mn.us/invasives/aquaticplants/purpleloosestrife/biocontrol.html. Accessed November 16, 2016.

Carson, Angela. "Purple Loosestrife: Sometimes Beauty Is a Beast." Dave's Garden. June 28, 2013. http://davesgarden.com/guides/articles/view/3157.

Connor, Larry. "August Crossroads." *Bee Culture: The Magazine of American Beekeeping*. July 25, 2016. http://www.beeculture.com/august-crossroads/.

Crone, Martha. "Purple Loosestrife." *Fringed Gentian* 6:2 (April 1958).

Eisterhold, Joe. "Purple Loosestrife (*Lythrum salicaria*) Management in Minnesota." Minnesota Department of Natural Resources. http://bugwoodcloud.org/mura/mipn/assets/File/MNWIISC%20talks/upload%20folder/Monday_PM_Management1_Eisterhold.pdf. Accessed November 16, 2016.

Encyclopedia of the Earth. "Purple Loosestrife." Online.

"Invasive Plants in Minnesota: Purple Loosestrife." Friends of the Wildflower Garden, Inc. 2013. http://www.friendsofthewildflowergarden.org/pages/plants/purpleloosestrife.html.

"How to Help Wildlife." *BBC Nature Features*. May 29,

2013. http://www.bbc.co.uk/nature/22433553.
Lavoie, Claude. "Should We Care about Purple Loosestrife?" *Biological Invasions* 12 (2010): 1967–99. doi:10.1007/s10530-009-9600-7.
"*Lythrum salicaria* 'Purple loosestrife.'" Seedaholic.com. 2016. http://www.seedaholic.com/lythrum-salicaria-purple-loosestrife.html.
Minnesota Sea Grant. "Purple Loosestrife: What You Should Know, What You Can Do." University of Minnesota. 2009. http://www.seagrant.umn.edu/ais/purpleloosestrife_info.
Moen, Sharon. "Where Have All the Purple Flowers Gone?" Minnesota Sea Grant. 2002. http://www.seagrant.umn.edu/newsletter/2002/09/where_have_all_the_purple_flowers_gone.html.
Pliny the Elder. *Naturalis Historia* (*Natural History*). Libri Vox. https://librivox.org/author/5293?primary_key=5293&search_category=author&search_page=1&search_form=get_results. Accessed November 16, 2016.
"Purple Loosestrife." Blue Line Honey: Helpful Information for Backyard Beekeepers. August 21, 2013. http://www.bluelinehoney.com/2013/08/.
"Purple Loosestrife (*Lythrum salicaria*)." Highbury Wildlife Garden. 2015. http://highburywildlifegarden.org.uk/the-garden/bees-faves/purple-loosestrife/.
"Purple Loosestrife (*Lythrum salicaria*)." Minnesota Department of Natural Resources. http://www.dnr.state.mn.us/invasives/aquaticplants/purpleloosestrife/index.html. Accessed November 16, 2016.
"Purple Loosestrife Management Program." Minnesota Department of Natural Resources. http://www.dnr.state.mn.us/invasives/aquaticplants/purpleloosestrife/program.html. Accessed November 16, 2016.
Rezanka, Richard, invasive species specialist, Minnesota Department of Natural Resources. Phone conversations. January 30, 2013, September 28, 2015.
Rindels, Sherry. "Purple Loosestrife." *Horticulture and Home Pest News*. Iowa State University Extension. May 5, 1994.
Shamsi, S. R. A., and F. H. Whitehead. "Comparative Eco-physiology of *Epilobium hirsutum L.* and *Lythrum salicaria* L. I.: General Biology, Distribution and Germination." *Journal of Ecology* 62 (1974): 272–90.
Skinner, Luke. "Something's Bugging Purple Loosestrife." *Minnesota Conservation Volunteer* (March–April 1998). http://www.dnr.state.mn.us/mcvmagazine/.
Skinner, Luke, former supervisor of the invasive species program, Minnesota Department of Natural Resources. Phone conversation. February 1, 2013.
Skinner, Luke C., William J. Rendall, and Ellen L. Fuge. "Minnesota's Purple Loosestrife Program: History, Findings, and Management Recommendations." Minnesota Department of Natural Resources. Special Publication 145. 1994. http://files.dnr.state.mn.us/publications/fisheries/special_reports/145.pdf.
USDA, Natural Resources Conservation Service. "Lythrum salicaria: Botanical and Ecological Characteristics." PLANTS Database. 2015. http://plants.usda.gov, http://www.fs.fed.us/database/feis/plants/forb/lytsal/all.html#BOTANICAL%20AND%20ECOLOGICAL%20CHARACTERISTICS.

7. SOYBEANS

"AgWeb Soybean Harvest News: 2015 AgWeb Soybean Harvest Map." AgWeb. 2015. http://www.agweb.com/crops/soybean-harvest-map/.
Bailey, Wayne, et al. "The Effectiveness of Neonicotinoid Seed Treatments in Soybeans." Joint publication of twelve US universities. December 2015. http://www.extension.umn.edu/agriculture/soybean/pest/docs/effectiveness-of-neonicotinoid-seed-treatments-in-soybean.pdf.
Bennett, J. Michael, Dale R. Hicks, and Seth L. Naeve. *The Minnesota Soybean Field Book*. University of Minnesota Extension. http://www.extension.umn.edu/agriculture/soybean/docs/minnesota-soybean-field-book.pdf. Accessed November 16, 2016.
Bowen, N. Dennis. "Soybean, Long Road to the Field." Crop, Stock and Ledger. University of Illinois Extension. http://web.extension.illinois.edu/cfiv/crop/090428.html. Accessed November 16, 2016.
"A Condensed History of Minnesota Agriculture, 1858–2008." Minnesota Department of Agriculture. 2008. http://www.mda.state.mn.us/news/publications/kids/maitc/sesquitimeline.pdf. Accessed 2014.
Daniels, Eve. "Soybean Futures." Global Food. MnDRIVE, Minnesota's Discovery, Research and Innovation

Economy. November 3, 2015. https://mndrive.umn.edu/food/news/soybean-futures.

Edwards, Ann Marie. "Soybean Breeding Improving Varieties for Minnesota." *Minnesota Farm Guide*. June 6, 2014. http://www.minnesotafarmguide.com/news/crop/soybean-breeding-improving-varieties-for-minnesota/article_c55c4d3e-e753-11e3-8627-001a4bcf887a.html.

"Field of Soybeans." University of Minnesota Agricultural Experiment Station. 2016. http://www.maes.umn.edu/publications/food-life/soybeans.

Food for Life. University of Minnesota Agricultural Experiment Station. 2001. http://www.maes.umn.edu/publications/food-life.

Gibson, Lance, and Garren Benson. "Origin, History, and Use of Soybeans (*Glycine max*)." Iowa State University. 2005. http://www.agron.iastate.edu/courses/agron212/Readings/Soy_history.htm.

Goldberg, Ray A. *The Soybean Industry, with Special Reference to the Competitive Position of the Minnesota Producer and Processor*. Minneapolis: University of Minnesota Press, 1952.

Gutknecht, Gilbert W. "Tribute to Jean W. Lambert." Address to the Minnesota House of Representatives. May 25, 2000.

"History of the Soybean." Minnesota Soybean Growers Association. 2009. http://www.mnsoybean.org/all-about-soy/soybeans/history-of-the-soybean/. Accessed 2014.

Krohn, Tim. "50 Years Ago at Mankato Soybean Plant, a Wall of Oil." *Mankato Free Press*, January 6, 2013. http://www.mankatofreepress.com/news/local_news/years-ago-at-mankato-soybean-plant-a-wall-of-oil/article_9dec7c89-958a-5e54-ac17-9f27a3773d61.html.

Lee, Stephen J. "Operation Save a Duck and the Legacy of Minnesota's 1962–63 Oil Spills." *Minnesota History* 58.2 (Summer 2002): 105–23. http://collections.mnhs.org/MNHistoryMagazine/articles/58/v58i02p105-123.pdf.

Meersman, Tom. "Report Says Insecticide-Coated Seeds Are Overused." *Star Tribune*, January 31, 2016, D1.

———. "Soybeans Planted on Record Acreage This Year in Minnesota." *Star Tribune*, June 30, 2015. http://www.startribune.com/soybeans-planted-on-record-acreage-this-year/311055001/.

"Minnesota Farmers Become Acquainted with Soybeans." University of Minnesota Extension. 2016. http://www.extension.umn.edu/about/history/1926-1950/minnesota-farmers-become-acqua/.

Mohr, Paula. "Checkoff Funds Boost Research." *Farm Progress* (August 2012): 8. http://magissues.farmprogress.com/TFM/TF08Aug12/tfm008.pdf.

———. "MSGA Celebrates 50 Years." *Farm Progress* (February 2012): 34. http://magissues.farmprogress.com/TFM/TF02Feb12/tfm034.pdf.

"One Last Showing for Dr. Orf at Northern Plot Tours." Minnesota Soybean Research and Promotion Council. September 1, 2015. https://mnsoybean.org/blog-msrpc/one-last-showing-for-dr-orf-at-northern-plot-tours/.

"Orf Leaves Indelible Mark on State's Soybean Industry." *Farm Progress* (April 2014): 26. http://magissues.farmprogress.com/TFM/TF04Apr14/tfm026.pdf.

Ruttan, Vernon W., George W. Norton, and Randy R. Schoeneck. "Soybean Yield Trends in Minnesota." *Minnesota Agricultural Economist* 621 (July 1980). University of Minnesota Agricultural Extension Service. http://ageconsearch.umn.edu/bitstream/163779/2/mn-ag-econ-621.pdf.

Shurtleff, William, and Akiko Aoyagi. "A Comprehensive History of Soy: History of Soybeans and Soyfoods Worldwide—Past, Present, and Future." Soyinfo Center. 2016. www.soyinfocenter.com/HSS/history.php.

———. "History of Soy in the United States, 1766–1900." From "History of Soybeans and Soyfoods, 1100 B.C. to the 1980s." Unpublished manuscript. Lafayette, CA: Soyfoods Center, 2004. http://www.soyinfocenter.com/HSS/usa.php.

Smith, Keith. "Soybean Meal Information Fact Sheet: Soybeans—History and Future." http://www.soymeal.org/FactSheets/HistorySoybeanUse.pdf. Accessed November 16, 2016.

"Soy Stats 2012: A Reference Guide to Important Soybean Facts and Figures." American Soybean Association. SoyStats.com.

"Soybean." Wikipedia. 2016. https://en.wikipedia.org/wiki/Soybean.

"Soybean Production." University of Minnesota Extension. http://www1.extension.umn.edu/agriculture/soybean/. Accessed November 16, 2016.

"Soyfoods Are Part of America's History." Soyfoods Association of North America. http://www.soyfoods.org/press-releases/soyfoods-are-part-of-americas-history. Accessed November 16, 2016.

USDA, National Agricultural Statistics Service. "Minnesota Agricultural Statistics." 2012. http://www.nass.usda.gov/Statistics_by_State/Minnesota/.

USDA, National Agricultural Statistics Service, Minnesota Field Office. "Annual Statistical Bulletin." 2015. http://www.nass.usda.gov/Statistics_by_State/Minnesota/Publications/Annual_Statistical_Bulletin/.

Vance, Daniel. "Lowell Andreas." *Connect Business* (July 2002). http://connectbiz.com/2002/07/lowell-andreas/.

8. WHEAT

Boehm, David, and Tracy Sayler. "Wheat Industry Yields Rich History." *Prairie Grains Magazine* (June 2002).

"Bonanza Farms." Minnesota Historical Society. http://www.mnhs.org/library/tips/history_topics/62bonanza.html. Accessed November 16, 2016.

"Bonanza Farms." Wikipedia. 2013. http://en.wikipedia.org/wiki/Bonanza_farms.

Cartwright, R. J. "Bruns and Finkle Grain Elevator, Moorhead." MNopedia. November 19, 2013. http://www.mnopedia.org/structure/bruns-and-finkle-grain-elevator-moorhead.

———. "Grasshopper Plagues, 1873–1877." MNopedia. April 29, 2015. http://www.mnopedia.org/event/grasshopper-plagues-1873-1877.

Danbom, David D. "Flour Power: The Significance of Flour Milling at the Falls." *Minnesota History* 58.5 (Spring/Summer 2003): 270–85. http://collections.mnhs.org/MNHistoryMagazine/articles/58/v58i05-06p270-285.pdf.

Dierckins, Tony. "Lost Landmark: Duluth's Imperial Mill." Zenith City Online. August 27, 2012. http://zenithcity.com/lost-landmark-duluths-imperial-mill/.

Easterbrook, Greg. "Forgotten Benefactor of Humanity." *The Atlantic*, January 1997. http://www.theatlantic.com/magazine/archive/1997/01/forgotten-benefactor-of-humanity/306101/.

"Flour Milling (1888–1957)." Zenith City Online. http://zenithcity.com/zenith-city-history-archives/duluth-industry/flour-milling-1888-1957/. Accessed November 16, 2016.

Fossum, Paul R. "Early Milling in the Cannon River Valley." Paper. Albert Lea session, ninth state historical convention. June 14, 1932.

Frame, Robert M. III. *Millers to the World: Minnesota's Nineteenth Century Water Power Flour Mills*. St. Paul: Minnesota Historical Society, 1977.

———. "Mills Machines and Millers: Minnesota Sources for Flour-Milling Research." *Minnesota History* 46.4 (Winter 1978): 152–62. http://collections.mnhs.org/MNHistoryMagazine/articles/46/v46i04p152-162.pdf.

———. "The Progressive Millers: A Cultural and Intellectual Portrait of the Flour Milling Industry, 1870–1930, Focusing on Minneapolis, Minnesota." Thesis, University of Minnesota. 1980.

Hazen, Theodore R. "Flour Milling in America: A General Overview." 2003. http://www.angelfire.com/folk/molinologist/america.html.

Holen, Doug. "Wheat Growing Makes Comeback in Minnesota." *West Central Tribune* (Willmar, MN), February 12, 2011. http://www.wctrib.com/content/wheat-growing-makes-comeback-minnesota.

Jaradat, Abdullah A. "Wheat Landraces: Genetic Resources for Sustenance and Sustainability." USDA Agricultural Research Service. https://www.ars.usda.gov/SP2UserFiles/Place/50600000/products-wheat/AAJ-Wheat%20Landraces.pdf. Accessed November 16, 2016.

Jarchow, Merrill E. "King Wheat." *Minnesota History* 29.1 (March 1948): 1–28. http://collections.mnhs.org/MNHistoryMagazine/articles/29/v29i01p001-028.pdf.

Johnson, Frederick L. "When Wheat Was King in Minnesota." *MinnPost*, September 10, 2013. https://www.minnpost.com/mnopedia/2013/09/when-wheat-was-king-minnesota.

Kelsey, Kerck. *Prairie Lightning: The Rise and Fall of William Drew Washburn*. Lakeville, MN: Pogo Press, 2010.

Kuhlmann, Charles B. "The Influence of the Minneapolis Flour Mills upon the Economic Development of Minnesota and the Northwest." *Minnesota History*

6.2 (June 1925): 141–54. http://collections.mnhs.org/MNHistoryMagazine/articles/6/v06i02p141-154.pdf.

Larson, Henrietta M. "The Wheat Market and the Farmer in Minnesota, 1858–1900." Thesis, Columbia University. 1926.

Lemke, Jeff. "Milling About the Train Yards: How Duluth's Railroads Served the Duluth Imperial Mill." Zenith City Online. July 24, 2014. http://zenithcity.com/milling-train-yards/.

"Norman Borlaug." Wikipedia. https://en.wikipedia.org/wiki/Norman_Borlaug. Accessed November 16, 2016.

Pennefeather, Shannon, ed. *Mill City: A Visual History of the Minneapolis Mill District*. St. Paul: Minnesota Historical Society Press, 2003.

"Pillsbury A-Mill." Wikipedia. 2015. https://en.wikipedia.org/wiki/Pillsbury_A-Mill. Accessed November 16, 2016.

"Resources: Wheat Facts." Wheat Foods Council. 2015. http://www.wheatfoods.org/resources/72.

Roberts, Kate. "Wheat." *Minnesota 150: The People, Places, and Things that Shape Our State*. St. Paul: Minnesota Historical Society Press, 2007.

Rogers, Col. George D. "History of Flour Manufacture in Minnesota." February 1905. https://archive.org/stream/historyofflourmaoorogerich/historyofflourmaoorogerich_djvu.txt.

Smalley, Eugene V. "The Flour Mills of Minneapolis." *Century Magazine* (May 1886).

Steen, Herman. *Flour Milling in America*. Minneapolis, MN: T. S. Denison and Company, 1963.

Storck, John, and Walter Dorwin Teague. *Flour for Man's Bread: A History of Milling*. Minneapolis: University of Minnesota Press, 1952.

Stuertz, Mark. "Green Giant." *Dallas Observer*, December 5, 2002. http://www.dallasobserver.com/news/green-giant-6389547.

USDA, National Agricultural Statistics Service. "All Wheat Planted Area." Map. 2016. http://www.nass.usda.gov/Charts_and_Maps/graphics/awacm.pdf.

Vietmeyer, Noel. *Our Daily Bread: The Essential Norman Borlaug*. Lorton, VA: Bracing Books, 2011.

Watts, Alison. "The Technology That Launched a City: Scientific and Technological Innovations in Flour Milling during the 1870s in Minneapolis." *Minnesota History* 57.2 (Summer 2000): 86–97. http://collections.mnhs.org/MNHistoryMagazine/articles/57/v57i02p086-097.pdf.

"Wheat." University of Minnesota Agricultural Experiment Station. 2015. http://www.maes.umn.edu/publications/food-life/wheat.

"Wheat Fields." http://www.millcitymuseum.org/wheat-fields. Mill City Museum, Minnesota Historical Society. Accessed November 16, 2016.

9. WHITE PINE

Bacig, Tom, and Fred Thompson. *Tall Timber: A Pictorial History of Logging in the Upper Midwest*. Minneapolis, MN: Voyageur Press, 1982.

Barzen, Mimi. "A History of Forestry in Minnesota." Minnesota Department of Natural Resources. 1969. http://files.dnr.state.mn.us/forestry/anniversary/documents/historyofForestry-1969.pdf.

Bell, Mary T. *Cutting Across Time: Logging, Rafting and Milling the Forests of Lake Superior*. Schroeder, MN: Schroeder Area Historical Society, 1999.

Birk, Douglas A. "Outta the Woods and Onto the Mills: Shifting Timber-Harvest Strategies on Minnesota's Early Lumbering Frontiers." From Site to Story: The Upper Mississippi's Buried Past. 1997. http://www.fromsitetostory.org/sources/papers/mnlogging/mnlogging2.asp.

Blegen, Theodore C. "With Ax and Saw: A History of Lumbering in Minnesota." *Forestry History Newsletter* 7.3 (Autumn 1963): 2–13.

Brady, Tim. "The Real Story of Chippewa National Forest." *Minnesota Conservation Volunteer* (November–December 2004). http://www.dnr.state.mn.us/mcvmagazine/.

Carroll, Francis M., and Franklin R. Raiter. *The Fires of Autumn: The Cloquet–Moose Lake Disaster of 1918*. St. Paul: Minnesota Historical Society Press, 1990.

Clatterbuck, Wayne K., and Leslie Ganus. "Tree Crops for Marginal Farmland: White Pine, with a Financial Analysis." Agricultural Extension Service, University of Tennessee. 2014. https://extension.tennessee.edu/publications/Documents/PB1462.pdf.

Conlin, Joseph R. "'Old boy, did you get enough of pie?' A Social History of Food in Logging Camps." *Journal of Forest History* (October 1979): 154–85.

Enger, Leif. "A History of Timbering in Minnesota." Minnesota Public Radio. November 16, 1998. http://news.minnesota.publicradio.org/features/199811/16_engerl_history-m/.

Flader, Susan L. *The Great Lakes Forest: An Environmental and Social History*. Minneapolis: University of Minnesota Press, 1983.

Forester, Jeff. *The Forest for the Trees: How Humans Shaped the North Woods*. St. Paul: Minnesota Historical Society Press, 2004.

Hoff, Mary. "A Century for the Trees." *Minnesota Conservation Volunteer* (March–April 2011). http://www.dnr.state.mn.us/mcvmagazine/.

Kominicki, John. "Frederick Weyerhaeuser Forged Strong Timber Roots." *Investor's Business Daily News*, March 19, 2014. investors.com/news/management/leaders-and-success/weyerhaeuser-collectivized-timber-and-created-sustainable-tree-farming.

Kraker, Dan. "Study Shows Deer Population Reducing Valuable North Shore Trees." Minnesota Public Radio. April 2, 2012. www.mprnews.org/story/2012/04/02/north-shore-deer-population.

Larson, Agnes M. *The White Pine Industry in Minnesota: A History*. Minneapolis: University of Minnesota Press, 2007.

"Lost Forty." Chippewa National Forest. http://www.fs.usda.gov/recarea/chippewa/recreation/recarea/?recid=26672&actid=70. Accessed November 16, 2016.

Minnesota Department of Natural Resources, Division of Forestry. "All about Minnesota Forests and Trees: A Primer." 1995, 2008. http://files.dnr.state.mn.us/forestry/education/primer/allaboutminnesota's forestsandtrees_aprimer.pdf.

———. *Connected to Our Roots: 100 Years of Growing Forests in Minnesota*. St. Paul: Minnesota DNR, 2010.

Minnesota Digital Library. "Minnesota's Logging History." Andersen Library, University of Minnesota. N.d.

"The Minnesota Pineries." *Harper's New Monthly Magazine* (March 1868). http://www.catskillarchive.com/rrextra/lgmnpi.Html.

"Minnesota's Forest Resources 2013." Minnesota Department of Natural Resources, Division of Forestry. July 2014. https://www.leg.state.mn.us/docs/2014/other/141117.pdf.

Peterson, Jim. "White Pine in Minnesota Forests." *Evergreen Magazine* (Spring 2000). http://wp_medialib.s3.amazonaws.com/wp-content/uploads/2014/10/EG_Spring2000.pdf.

Rajala, Jack. *Bringing Back the White Pine*. Duluth, MN: Pro Print, 1998.

Rice, Anna. "General Christopher C. Andrews: Leading the Minnesota Forestry Revolution." *History Teacher* 36.1 (November 2002): 91–115.

Ritter, Luke. "Frederick Weyerhaeuser (1834–1914)." Immigrant Entrepreneurship. May 27, 2015. http://immigrantentrepreneurship.org/entry.php?rec=239.

Robertson, Tim. "Red Lake Band to Reforest 50,000 Acres." Minnesota Public Radio. August 25, 2011. http://minnesota.publicradio.org/display/web/2011/08/24/red-lake-band-reforest-50000-acres.

Stearns, Forest W. "History of the Lake States Forests: Natural and Human Impacts." Great Lakes Ecological Assessments Reports. 1997. http://www.ncrs.fs.fed.us/gla/reports/history.htm.

"Weyerhaeuser, Frederick (1834–1914)." German Corner. 1996. http://www.germanheritage.com/biographies/mtoz/weyerthaeuser.html.

The White Pine Society. Wildlife Research Institute. http://whitepines.org/. Accessed November 16, 2016.

10. WILD RICE

Anders, Jake, Rose Carlson, Sarah Niskenen, Janell Stauff, and Josh Weise. "Genetic Research on Wild Rice and Its Cultural Implications." Report for Anthropology Senior Seminar, University of Minnesota, Duluth. May 6, 2004. http://www.d.umn.edu/~ande2927/Report.pdf.

Brody, Jane. "Scientists Tame Wild Rice, the 'Caviar of Grains.'" *New York Times*, November 25, 1986.

Carver, Jonathan. *Travels through the Internal Parts of North America in the Years 1766, 1767, 1768*. London: The author, 1780.

Clark, Ed. "Where the Wild Things Grow." AgWeb. January 11, 2012. http://www.agweb.com/article/where_the_wild_things_grow/.

Dunbar, Elizabeth, and Tom Scheck. "MPCA Seeks Lake-by-Lake Plan to Protect Wild Rice." Minnesota Public Radio. March 25, 2015. http://www.mprnews.org/story/2015/03/25/mpca-wild-rice.

Edman, F. Robert. *A Study of Wild Rice in Minnesota*. St. Paul: Minnesota Resources Commission, Upper Great Lakes Regional Commission, 1969.

Eisenthal, Jonathan. "Unlocking Wild Rice's Health Benefits." *Ag Innovation News* 22.1 (January–March 2013): 3. Agricultural Utilization Research Institute. http://www.auri.org/2013/01/unlocking-wild-rices-health-benefits/.

Gunderson, Dan. "Ojibwe Rice Harvest Is Latest Test of Treaty Limits in Minnesota." Minnesota Public Radio. August, 22, 2015. http://www.mprnews.org/story/2015/08/27/ojibwe-rice-harvest.

Gunderson, Dan, and Chris Julin. "Wild Rice at the Center of a Cultural Dispute." Minnesota Public Radio. September 24, 2012. http://news.minnesota.publicradio.org/features/200209/22_gundersond_wildrice-m/.

Hemphill, Stephanie. "Current Sulfate Standard Is about Right to Protect Wild Rice, Research Indicates." *Minn Post*, February 26, 2014. https://www.minnpost.com/environment/2014/02/current-sulfate-standard-about-right-protect-wild-rice-research-indicates.

Hudson, Bill. "Chippewa Tribes to Challenge Rights to Off-Reservation Ricing." WCCO. August 20, 2015. http://minnesota.cbslocal.com/2015/08/20/chippewa-tribes-to-challenge-rights-to-off-reservation-ricing/.

Jenks, Albert Ernest. *The Wild Rice Gatherers of the Upper Lakes: A Study in American Primitive Economics*. Washington, DC: Government Printing Office, 1901. https://archive.org/details/wildricegatheree00jenk.

Kraker, Dan, "Changing the Protection of Wild Rice Water: What You Need to Know." Minnesota Public Radio. October 27, 2015. http://www.mprnews.org/story/2015/10/27/wild-rice-water-standard.

———. "Fates of Wild Rice, Mines Intertwined in Northern Minnesota." Minnesota Public Radio. January 9, 2014. http://www.mprnews.org/story/2014/01/04/fates-of-wild-rice-mining-intertwined-in-northern-minn.

LaDuke, Winona. "Ricekeepers." *Orion Magazine* (July 2007). https://orionmagazine.org/article/ricekeepers/.

———. "Wild Rice and Ethics." *Cultural Survival Quarterly* (Fall 2004). https://www.culturalsurvival.org/publications/cultural-survival-quarterly/united-states/wild-rice-and-ethics.

———. "Wild Rice and Genetic Research/Modification, MN, 2006." The Pluralism Project. Harvard University.

Lofstrom, Ted. "The Rise of Wild Rice Exploitation and Its Implications for Population Size and Social Organization in Minnesota Woodland Period Cultures." *Minnesota Archaeologist* 46.2 (1987): 3–15.

Minnesota Conservation Volunteer. "Wild Rice from Harvest to Home." Video. September 16, 2016. http://www.youtube.com/watch?v=qAYlYmcgGZU.

Minnesota Department of Natural Resources. "Natural Wild Rice in Minnesota: A Wild Rice Study Document Submitted to the Minnesota Legislature by the Minnesota DNR." February 15, 2008. http://archive.leg.state.mn.us/docs/2008/mandated/080235.pdf.

Moyle, John B. "Wild Rice in Minnesota." *Journal of Wildlife Management* 8.3 (July 1944): 177–84.

Nelson, Beth, Minnesota Cultivated Wild Rice Council. Phone conversation. March 21, 2016.

Nelson, Jennifer J. "Crop Profile for Wild Rice in Minnesota." Minnesota Pesticide Impact Assessment Program (PIAP). January 1, 2000. http://www.ipmcenters.org/cropprofiles/docs/mnwildrice.pdf.

Nibi and Manoomin Committee 2015. "Native Wild Rice." Nibi Manoomin Symposium, Mille Lacs, MN. September 28–29, 2015. http://www.nativewildricecoalition.com/.

Oelke, Ervin. *Saga of the Grain: A Tribute to Minnesota Cultivated Wild Rice Growers*. Lakeville, MN: Hobart Publications, 2007.

"Protecting Minnesota's Wild Rice Lakes: June 2015 Snapshots." Minnesota Board of Water and Soil Resources. 2015. http://www.bwsr.state.mn.us/news/webnews/june2015/2.pdf.

"Treaties, Law, and Case Law." Protect Our Manoomin. http://www.protectourmanoomin.org/treaties-law-and-case-law.html. Accessed November 16, 2016.

USDA, National Institute of Food and Agriculture. "Native Wild Rice Coalition." http://www.nativewildricecoalition.com/. Accessed November 16, 2016.

Vennum, Thomas, Jr. *Wild Rice and the Ojibway People*. St. Paul: Minnesota Historical Society Press, 1988.

Walker, Deborah. *The Good Life: Mino-Bimadiziwin*.

RedEye Video. 1997. http://www.folkstreams.net/film,218.
White Earth Land Recovery Project (WELRP). "Comments on White Earth Land Recovery Project." September 10, 2013. http://welrp.org/.
"Wild Rice." 1854 Treaty Authority, Grand Portage, Bois Forte. 2016. http://www.1854treatyauthority.org/wildrice/.
"Wild Rice." University of Minnesota Agricultural Experiment Station. 2016. http://www.maes.umn.edu/publications/food-life/wildrice.
"Wild Rice, Current Research." University of Minnesota, North Central Research and Outreach Center. 2013. http://ncroc.cfans.umn.edu/research/wildrice.
"Wild Rice: Ecology, Harvest, Management." University of Wisconsin Extension. 2007. http://www.uwex.edu/ces/regionalwaterquality/Publications/WildRiceBrochure.pdf.
"Wild Rice and the Sulfate Standard." WaterLegacy. March 2011. http://waterlegacy.org/sites/default/files/Media/WildRiceSulfateFacts.pdf.
"Wild Rice Belies Its Name." Vegetarians in Paradise. 2013. http://www.vegparadise.com/highestperch311.html.
"Wild Rice Management." Minnesota Department of Natural Resources. 2013. http://www.dnr.state.mn.us/wildlife/shallowlakes/wildrice.html.
Yakowicz, Susie. "Manoomin, Sacred Grain of the Chippewa." Native American/First Nations History. September 24, 2009. http://suite101.com/article/manoomin-sacred-grain-of-the-chippewa-a152131. Accessed 2014.

CONCLUSION

Hawken, Paul. *The Ecology of Commerce: A Declaration of Sustainability*. New York: Harper Collins Publishers, 1993.

INDEX

FURTHER DETAILS ON SELECT IMAGES

page ii–iii Shutterstock, ©kzww

MNHS COLLECTIONS

page 4	Original Vegetation of Minnesota	G4141 .D2 1930 .M3 c.7 8F
page 15	Grimms	por 24279 r1
page 39	St. Paul house	MR2.9 SP3.2q r150
page 52	John Harris	por 24743 r1
page 52	apple orchard	SA2.3 p56
page 55	Peter Gideon	por 19936 p1
page 69	Gail Page	MC4.9 MV9 p3
page 75	protecting corn	E91.32 r10
page 79	harvesting corn	2119-B
page 84	canning corn	HD7.2 p32
page 89	Corn and Potato Palace	MA6.9 AN9 r2
page 97	sitting on lawn	6696-A
page 99	mowing	GV8.2 p139
page 150	Pickwick Mill	MW7.9 P r9
page 158	Duluth Imperial trade card	2007.64.986
page 159	Pillsbury A Mill	MH5.9 MP3.1P r97
page 160	Miss Minnesota	SA2.6 r1
page 167	Minnesota Harvest Field	AV1983.221
page 174	man with board	RUNK 218
page 176	bunkhouse near Bemidji	HD5.7 r37
page 177	log drive	HD5.41 r59
page 179	logjam on St. Croix	HD5.44 p7
page 180	Frederick Weyerhaeuser	AV1992.37.13
page 196	woven cedar bag	6874.11

LIBRARY OF CONGRESS

page 80	World War I poster	LC-USZC4-10124
page 168	dust storm	LC-USF34- 001615-ZE

NATIONAL ARCHIVES

page 135	World War II poster	44-PA-2525
page 200	parching rice	NRE-75-COCH(PHO)-1494
page 201	winnowing rice	NRE-75-COCH(PHO)-621

INSTITUTE FOR REGIONAL STUDIES

page 155	Dalrymple bonanza farm	G570.B874

Ten Plants That Changed Minnesota was designed
and set in type by Judy Gilats. The text face is Fairplex.
Champion and Acumin were used for sidebars and heads.
The book was printed by Versa Press in Peoria, Illinois.